P9-CJT-083

WITHDRAWN
UTSA LIBRARIES

Animal Cell Technology:
Principles and products

The Biotechnology Series

This series is designed to give undergraduates, graduates and practising scientists access to the many related disciplines in this fast developing area. It provides understanding both of the basic principles and of the industrial applications of biotechnology. By covering individual subjects in separate volumes a thorough and straightforward introduction to each field is provided for people of differing backgrounds.

Titles in the Series

Biotechnology: The Biological Principles M.D. Trevan, S. Boffey, K.H. Goulding and P. Stanbury
Fermentation Kinetics and Modelling C.G. Sinclair and B. Kristiansen (Ed. J.D. Bu'lock)
Enzyme Technology P. Gacesa and J. Hubble
Animal Cell Technology: Principles and Products M. Butler

Upcoming Titles

Monoclonal Antibodies
Biosensors
Industrial Fermentation
Plant Cell and Tissue Culture
Biotechnology Safety
Chemical Engineering for Biotechnology
Bioelectronics
Bioreactors

Overall Series Editor

Professor J.F. Kennedy *Birmingham University, England*

Series Editors

Professor J.A. Bryant *Exeter University, England*
Dr R.N. Greenshields *Biotechnology Centre, Wales*
Dr C.H. Self *Hammersmith Hospital, London, England*

The Institute of Biology IOB

This series has been editorially approved by the ***Institute of Biology*** *in London. The Institute is the professional body representing biologists in the UK. It sets standards, promotes education and training, conducts examinations, organizes local and national meetings, and publishes the journals* ***Biologist*** *and* ***Journal of Biological Education.***

For details about Institute membership write to: Institute of Biology, 20 Queensberry Place, London SW7 2DZ.

Animal Cell Technology: Principles and Products

M. Butler

Co-published by:
Open University Press
Open University Educational Enterprises Limited
12 Cofferidge Close
Stony Stratford
Milton Keynes MK11 1BY, England

Taylor and Francis
Publishing office
3 East 44th Street
New York NY 10017
USA

Sales office
242 Cherry Street
Philadelphia, PA 19106
USA

First Published 1987

British Library Cataloguing in Publication Data
Butler, Michael
Animal cell technology: principles and products.
1. Biotechnology 2. Cell culture
I. Title
660'.6 TP248.25.C4

ISBN 0-335-15169-8

ISBN 0-335-15167-1 Pbk

Library of Congress Cataloging in Publication Data
Butler, Michael, 1928–
Animal cell technology.

Bibliography: p.
Includes index.
1. Animal cell biotechnology. I. Title.
TP248.27.A53B88 1987 615'.36 87-10160
ISBN 0-8448-1517-9

ISBN 0-8448-1518-7 (pbk)

Project Management: Clarke Williams
Printed in Great Britain

Contents

Preface and Acknowledgements

Over the past 5 to 10 years, biotechnology has been introduced as an integral component of most degree-level courses in the biological sciences. Biotechnology is a broad based subject and its component parts fit into a range of traditionally accepted biology units — biochemistry, microbiology and immunology.

Animal cell technology is one such component. It is a technology which has developed from years of cell culture study on a small laboratory scale (~100 ml). Such cell cultures have been used to investigate some of the fundamental areas of biology — cellular metabolism, growth and the action of hormones. Such small-scale cultures have been found increasingly useful in toxicological testing, and where possible being preferred to whole animal experiments.

More recent developments have allowed animal cells to be cultured at a larger scale (100–10000 l) for the production of an increasing range of biological compounds. They are high value compounds with medical and veterinary applications — for diagnosis, prophylaxis or therapy. They have a high specific activity *in vivo* and enable appreciable metabolic changes to result from the clinical administration of relatively low doses (~ng/ml of blood plasma) which can be several orders of magnitude lower than those of bacterial therapeutics such as antibiotics. The commercial production of these animal cell biologicals normally requires cell culture at a scale of 100–1000 litres — animal cell cultures above this volume being quite exceptional.

Having identified an animal cell product of commercial value, there are a number of possible methods for production. The first to be considered may be biochemical extraction from bulk biological tissue. However, this method has a number of disadvantages — it is unreliable, suffers from batch variation and often insufficient tissue is available to satisfy the demand for the final purified product. Alternative approaches include chemical synthesis, animal cell culture or culture of genetically engineered bacteria. For each product the preferred approach may be different. Chemical synthesis may suit relatively small molecules. Production from genetically engineered bacteria may be favoured for proteins which are normally unmodified by eucaryotic post-translational events, or where activity is unaffected by glycosylation. Although once considered difficult, animal cell cultures can now produce routinely acceptable yields of complex products. The most common in production are the viral vaccines, marking the beginning of many of the now established principles of animal cell technology.

The first part of this book describes the basic principles involved in the use of animal cell cultures, the possibilities for genetic modification and the problems to consider in scaling up such cultures. In the scond part of the book, several selected cell products are chosen as case studies for considering alternative strategies for large-scale production. The background of our understanding of these cell products is discussed and related to the requirements of large-scale production.

Clearly, each cell product will have an associated set of arguments favouring one or other process for its large-scale production. In many instances such production involves genetic engineering in bacteria, and no excuse is offered for introducing these techniques in a book on animal cell technology. Future developments may change the preferred production methods for some of these cell products. One particular prediction may be the increased use of genetically engineered animal cells, the large-scale culture of which may allow production processes which express the advantages (but few of the disadvantages) of using genetically engineered bacteria.

The products selected for discussion in this book are those that have become prominent in most people's minds when considering large-scale process production of animal cell products. Clearly, the full list of commercially valuable cell products is much longer than those described and many more novel products are expected in the near future. The development of production processes for these new compounds will undoubtedly be promoted by the on-going discussion concerning production of those more well established biologicals which are described in this book.

I am grateful to those many students who unwittingly became guinea pigs to the development of some of the thoughts and ideas in this book. Although the work is primarily aimed at the level of final year degree students, it may also be of benefit to post-graduate or other research workers when first entering the maze of animal cell technology. It assumes a basic knowledge of cell biology and biochemistry which most biological science students will study in the initial years of their course. Hopefully this book will help in solving some of the existing set of problems so as to continue the rapid advance of animal cell technology.

I would like to express my gratitude to my colleagues at Manchester Polytechnic, particularly Dr Maureen Dawson and Dr Peter Gowland for their valuable comments and criticism during the preparation of the manuscript. Also, my thanks go to Professor Bryan Griffiths for his thorough review and comments before publication and to my wife, Elisabeth for her patience during many hours of proof reading.

Figure and Table Acknowledgements

Chapter 1

Fig. 1.3 Hayflick, L. and Moorhead, P.S. (1961) *Exp. Cell Res.* **25**, pp. 585-621.

Chapter 2

Fig. 2.6 Butler, M. *et al.* (1983) *J. Cell Sci.* **61**, pp. 351-363.
Fig. 2.7 Kluft, C. *et al.* (1983) *Adv. Biotech. Processes* **2**, 98-109.
Fig. 2.9 Birch, J.R. *et al.* (1987) in 'Plants and Animal cells: Process Possibilities' (Webb, C. and Mavituna, F. Eds.) pp. 162-171. Chichester, Ellis Horwood.
Table 2.2 Katinger, H. and Scheirer, W. (1985) in 'Animal Cell Biotechnology' vol. 1 (Spier, R.E. and Griffiths, B. Eds.) pp. 167-193 London, Academic Press.
Table 2.3 Griffiths, B. (1986) *Trends in Biotech.* **4**, pp. 268-272.

Chapter 3

Table 3.2 Petricciani, J.C. (1985) in 'Large-scale Mammalian Cell Culture' (Feder, J.B. and Tolbert, W.R. Eds.) London, Academic Press.

Chapter 5

Fig. 5.1 Sternberg, M.J.E. and Cohen, F.E. (1982) *Int. J. Biol. Macromol.* **4**, pp. 137-144.
Fig. 5.3 and 5.4 Friedman, R.M. (1981) 'Interferons — A Primer' pp. 33-37. New York, Academic Press.

Chapter 6

Fig. 6.4 and 6.5 Birch, J.R. *et al.* (1985) *Trends in Biotechnology* **3**, p. 162.
Fig. 6.7 and 6.8 Campbell, A.M. (1984) *Monoclonal Antibody Technology*, pp. 42-43.
Table 6.2 Reuveny, S. *et al.* (1985) *Develop. Biol. Stand.* **60**, p. 185.

Chapter 9

Fig. 9.3 Klausner, A. (1986) *Biotech.* **4**, pp. 706-710.
Fig. 9.5 Kluft, C. *et al.* (1983) *Adv. in Biotech. Processes* **2**, pp. 97-110.

SECTION I

Animal Cell Technology: Principles

Chapter 1

The Development of Cell Cultures

Historical Background

The first recorded attempts to maintain animal cells in culture can be attributed to Ross Harrison, who in 1907 devised the hanging drop technique. He suspended dissected nerve tissue from frog embryos in lymph fluid which was allowed to clot as a droplet on the underside of a microscope cover slip. Figure 1.1 shows how this was mounted on a hollow microscope slide and sealed with wax. By this method Harrison was able to observe growth of the embryonic nerve cells for several weeks.

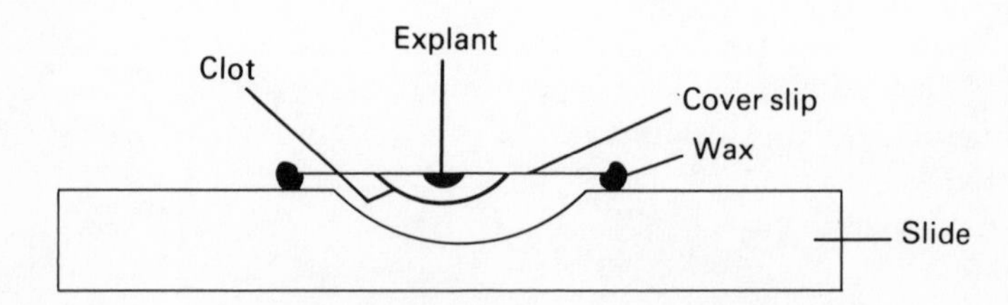

Fig. 1.1 Harrison's hanging drop technique (1907).

This work was extended by Burrows, who in collaboration with Alexis Carrel established techniques for the culture of a wide range of mammalian cells. They showed that strict aseptic control could enable the prolonged sub-culture of these cells for several years without infection. Good survival and growth of the cells was attained by the use of chick embryo extracts which were mixed with the plasma. The nutrients and growth factors were provided by the embryo extracts and the substratum for cell attachment was provided by the matrix of the fibrin formed from the plasma. Cultures were maintained in Carrel flasks which were small flat-bottomed containers devised for ease of aseptic manipulation.

The use of trypsin, introduced by Rous for the dispersion of cells from tissue explants, was a major innovation in obtaining cell suspensions. This allowed cultures to be established from single cell types and distinguished the technique of 'cell culture' from 'tissue culture'. Although the terms are often (but incorrectly) interchanged, tissue culture involves the growth of cells *in vitro* in a tissue matrix which may involve several cell types, whereas cell culture involves the growth of cells as independent micro-organisms.

The use of trypsin became particularly valuable in establishing procedures for sub-culturing cells. However, before the 1940s the widespread use of culture techniques was still limited by the stringent sterility controls necessary. The perception of these difficulties was also intensified by Carrel's fastidious nature in insisting on full surgical dress akin to those used in hospital operating theatres. It was not until the late 1940s, with the advent of antibiotics, that further developments of cell culture techniques took place. The addition of antibiotics (such as penicillin and streptomycin) to culture media eased the handling of cultures by reducing the chances of bacterial infections — a particular problem with the chemically undefined biological fluids and extracts used at the time.

The use of an antibiotic-supplemented culture media which was less prone to infection established cell culture as a routine technique which could be developed further. During this period human carcinoma cells such as the well known HeLa cell line were isolated and found to proliferate vigorously in culture. Large-scale animal cell cultures were first considered after Enders' discovery in the late 1940s that viruses could be propagated in cell cultures and used as vaccines. The consequent development of cell culture scale-up for polio vaccine production in the 1950s heralded the beginning of animal cell culture as a developing technology.

In the early 1950s, Earle and Eagle made a significant contribution to this technology by their analysis of the nutritional requirements of cultured cells. In 1955 the media formulation known as Eagle's minimum essential medium (EMEM — Table 1.1) defined the growth requirements of HeLa cells and mouse *L* cells in terms of optimal concentrations of chemically defined nutrients. This formulation was based on the compounds required to replace the growth promoting biological fluids previously used for cell growth. This chemically defined media had the advantages of consistency between batches, ease of sterilization and lowered chance of contamination. EMEM and its subsequent various modifications have been used extensively for cell culture. However, these formulations do require an additional supplement (usually 10%) of chemically undefined blood serum to provide unidentified growth factors and hormones.

Since these developments in the 1950s the range of available cell lines has increased considerably. Many of the cells have undergone modifications from their original *in vivo* state. This has involved the transformation to continuous cell lines capable of infinite growth capacity, or in some cases their genetic manipulation to produce selected products. In 1975 Kohler and Milstein exploited the ability to

Table 1.1 Eagle's Minimum Essential Medium (MEM)

L-Amino acids	*mM*	*Vitamins/Co-factors*	*μM*
Arginine	0.6	Choline	8.3
Cystine	0.1	Folic acid	2.3
Glutamine	2.0	Inositol	11.0
Histidine	0.2	Nicotinamide	8.2
Isoleucine	0.4	Pantothenate	4.6
Leucine	0.4	Pyridoxal	6.0
Lysine	0.4	Riboflavin	0.27
Methionine	0.1	Thiamine	3.0
Phenylalanine	0.2		
Threonine	0.4	*Inorganic ions*	mM
Tryptophan	0.05	NaCl	116
Tyrosine	0.2	KCl	5.4
Valine	0.4	$CaCl_2$	1.8
		$MgCl_2.6H_2O$	1.0
Glucose	5.5	$NaH_2PO_4.2H_2O$	1.1
		$NaHCO_3$	23.8
Supplemented Serum	10%		
		Phenol red	5 mgl^{-1}

fuse cells of different types to produce genetically stable hybridomas capable of continuous secretion of pre-determined monoclonal antibodies. The range and potential of these cell hybrids is still being explored.

The era of recombinant DNA technology developed in the 1970s with the ability to express mammalian genes in bacteria, and this allowed the production of a range of useful mammalian proteins in bacterial cultures. These proteins included insulin, somatotrophin (growth hormone), interferon and many others which are discussed in later chapters. Although it was once considered that the use of such recombinant bacteria might replace the need for animal cell cultures, it is now clear that both systems have their own merits and demerits which should be considered for the commercial production of any biological compound. Now, a range of recombinant mammalian cells have been developed with capabilities of high specific production of selected compounds.

The range of commercially useful biologicals produced from animal cell cultures is now expanding. Most are proteins or glycoproteins and are of use as pharmaceutical, medical or veterinary products. There are available typically as extracellularly-released products of an expanding variety of cell lines which have been induced into culture. However, in order to supply would-be markets they are needed in relatively large quantities, and this requires a careful study and optimization of the production processes involved. Such a study should start with an examination of the principles of cell biology and biochemical engineering that underlie such production processes.

Characteristics of Mammalian Cells

Animal cells can be derived from tissue explants of four basic types — epithelial, connective, muscle or nervous tissues. Some of these cells, such as the haemopoietic cells of the bone marrow or lymphocytes derived from the lymphoid tissue (broadly classified as connective), are found in suspension in body fluids and are non-anchorage-dependent when grown in culture. Lymphocytes in particular have been well studied. Under microscopic examination they appear spherical, with a diameter of 10–20 μm, and they can be induced to grow in a liquid suspension culture in a similar way to bacteria if the appropriate nutrients and growth factors are provided.

All other normal mammalian cells which are not derived from blood or lymph fluids are anchorage-dependent, i.e. they require a surface substratum for attachment and growth. This characteristic clearly distinguishes animal cells from bacteria and has an important bearing on culture systems. The most widely used anchorage-dependent cell types obtained from tissues of normal animals are epithelial or fibroblast cells. When attached to a surface they form a single cell layer (monolayer) and can be recognized by their characteristic appearance under microscopic examination as cobble-stone or spindle-shaped respectively (Fig. 1.2).

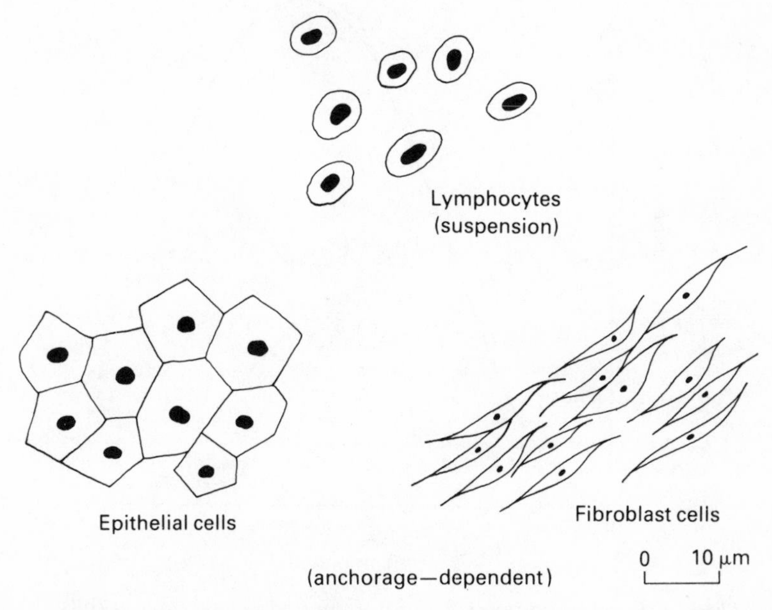

Fig. 1.2 Morphology of 3 cell types commonly used in culture.

Cells which are taken directly from an animal or human tissue form a primary culture *in vitro*. The excised tissue is generally macerated physically before treatment with a proteolytic enzyme — normally trypsin. After washing and removal of cell debris the isolated cells are suspended in growth medium to form the primary culture. A secondary culture can be formed by sub-culture from this primary culture. This generally involves removal of the growth media and washing the cell monolayer before trypsin treatment to detach the cells from their substratum. The cells are then treated with an anti-trypsin (such as serum) before washing and inoculation into fresh growth medium. A cell line is established by repeated culture through cycles of growth, trypsinization and sub-culture. For some cells this process may be continued indefinitely whereas in other cases there is a finite growth capacity.

Two cell lines isolated from normal, non-cancerous human embryonic tissue — the MRC-5 and WI-38 cells — have proved particularly valuable for experimental work and use in large-scale cultures. The lines are designated by the initials of the establishments in which they were first isolated — Medical Research Council and Wistar Institute. Such cells have been used extensively in the

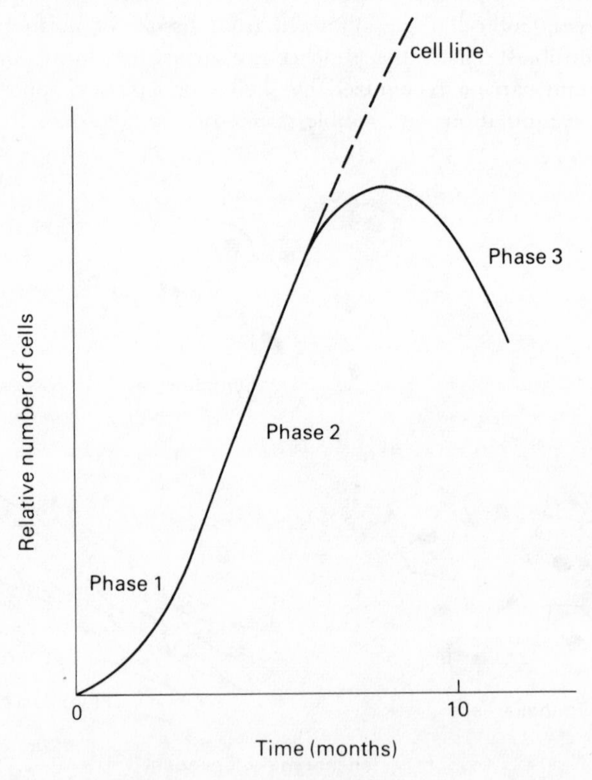

Fig. 1.3 Growth potential of human embryonic diploid cells.
(From Hayflick, 1961)

commercial production of human therapeutic agents — particularly viral vaccines. They are stable, free of contaminating agents and can easily be monitored for 'normality'.

In 1961 Hayflick and Moorhead showed that these cells were capable of extensive but finite growth over several months through three distinctive phases (Fig. 1.3). Initial slow growth (Phase 1) is followed by a period of exponential growth, in which the cells may be sub-cultured for about 50 generations, after which time the cells enter a senescent phase in which there is an observed decline in viable cell numbers. The finite growth capacity of these cells suggests that they go through an intrinsic ageing process. Cells can be stored in freezing solutions in liquid nitrogen at any point during the growth curve, and subsequent thawing will allow the cells to continue growth from the same point of the curve (Fig. 1.3). Despite the finite life span of these human diploid cells, their growth potential is sufficient to provide a substantial cell biomass which is more than adequate for the needs of commercial production of vaccines (see calculation in Table 1.2).

Cells associated with tumours are readily established in culture because of their good growth characteristics. The most widely used human cell line of this type is the HeLa line derived from a cervical cancer from a patient in 1952 — Henrietta Lach. The genetic complement of these cells is abnormal and described as hypotetraploid — each cell containing about four copies of each chromosome. However, the cells are particularly robust and have high growth rates in suspension culture. They have been extensively cultured *in vitro* for several decades and their growth capacity appears unlimited.

Table 1.2 Cellular biomass produced over 50 generations of cell growth

1 Exponential growth of cells can be represented by

$$N = N_o.2^x$$
$$\text{or } \log N = \log N_o + x.\log 2, \qquad (1)$$

where N = final cell number
N_o = initial cell number
x = number of generations of exponential growth.

2 Consider an initial cell number (N_o) of 10^7 which would probably be contained in 10 mm^3 of tissue.

3 Over 50 generations the cell number (N) will increase

$$\log N = 7 + 50.\log 2 = 22$$
$$N = 10^{22}.$$

Such a cell number would be accommodated in 10^9 m^3 of tissue

(equivalent to the biomass of 100 million people) or 10^{13} l of culture.

Table 1.3 Characteristics of 'normal' primary cells

diploid
finite life span
anchorage-dependent
density inhibited
non-malignant

The transformation of normal animal cells involves a change which allows the formation of a continuous cell line. Such transformed cell lines can continue exponential growth indefinitely and follow the extrapolated line of Phase 2 shown in Fig. 1.3. Transformation may occur spontaneously (as is often found in rodent cells) or may be induced by application of carcinogens or selected viruses to the cultured cells. The change to infinite growth capacity may or may not be associated with changes of some of the other characteristics of 'normal' cells as listed in Table 1.3. Frequently transformation causes changes in the chromosome complement, resulting in aneuploid or heteroploid cells — cells containing chromosome numbers other than the diploid number. The transformed cells may be induced to grow in suspension but some remain preferentially anchorage-dependent. Although the molecular basis of cell transformation is not fully understood, the cellular changes involved are undoubtedly similar to those associated with tumorigenesis in live animals. However, not all cell lines transformed in culture are malignant and this is shown by their inability to form tumours in experimental animals.

It was decided in the early days of vaccine production that transformed cells would be unsuitable for the production of human injectable products (such as vaccines) because of the perceived dangers of transmitting carcinogenic agents to the final products. The safety criteria adopted at this time only allowed the use of primary cells taken from a healthy animal, although some time later 'normal' human diploid cells such as WI-38 cells were accepted. The designation of cells as 'normal' is based on a set of characteristics outlined in Table 1.3 and include:

(1) A diploid chromosomal complement, which indicates that no gross genetic changes have occurred
(2) Anchorage-dependence and density inhibition, which is an indication of growth control, and is shown by cessation of growth as a confluent monolayer on the surface of a growth flask
(3) A finite life span, which is a reflection of the intrinsic growth regulation of the cells
(4) Non-malignancy of the cells, which can be shown by the inability to form tumours following injection of the cells into immuno-compromised mice

In order to inoculate cultures conveniently, it is of considerable advantage be able to store. cells. This allows the maintenance of cell stocks for culture inocula without having to resort to primary animal tissue. Such a master cell stock is important so that:

(1) Any genetic changes resulting from continuous sub-culture of a cell line can be monitored by comparison with the original cells
(2) A cell line is not lost if a particular culture becomes contaminated

These stocks are maintained by storage of the cells under liquid nitrogen at – 196°C. However, freezing and thawing of cells in an aqueous culture medium can cause sufficient cell damage through ice crystal formation to result in complete loss of viability. To avoid such problems, cryopreservatives such as dimethyl sulphoxide (DMSO) or glycerol can be used to protect cells against ice damage and enable cells to be stored in liquid nitrogen for indefinite periods of time. The existence of large international cell repositories enables long-term storage of a wide variety of cell lines for reference and purchase as needed.

Transformed cells can be maintained indefinitely through cycles of culture and storage. Diploid cells with a finite life span will retain a residual growth capacity after storage, dependent upon the number of previous generations of growth. Thus, human diploid cells with a total growth capacity of 50 generations and grown for 20 generations retain a capacity for a further 30 generations of growth even after storage.

Cell Growth and Culture Conditions

The growth of animal cell cultures follows a pattern similar to the kinetics widely studied in bacteria (Fig. 1.4). The initial cell inoculum requires a period of adaptation to fresh medium and this will be shown as a lag phase. The length of this phase will depend upon the previous state of the cells. Sub-culture of growing cells into fresh media at around 10^5 cells/ml will require little adaptation and the lag phase may be absent. However, transfer from an alternative medium or from liquid nitrogen storage may prolong this phase.

The exponential growth phase proceeds according to equation (1) in Table 1.2 with a typical doubling time of 18–24 h for healthy well adapted cells, although longer doubling times may be experienced with some fastidious or problematic cells. These values should be compared with those of bacteria whose doubling times can be as low as 15 min. This indicates the problem of bacterial contamination which can very quickly outgrow the mammalian cells in culture.

Cell growth can continue in culture until some growth limitation is encountered. For anchorage-dependent cells this will often be the available surface area. Once a monolayer of cells is formed over this growth surface, contact between individual cells can inhibit further cell division. Nutrient limitation or inhibitor accumulation are other possibilities for arresting exponential growth. At this point the cells are in a stationary phase, the length of which is dependent upon the nature of the growth limiting factor. Eventually, a death phase results in a decline in total number of viable cells.

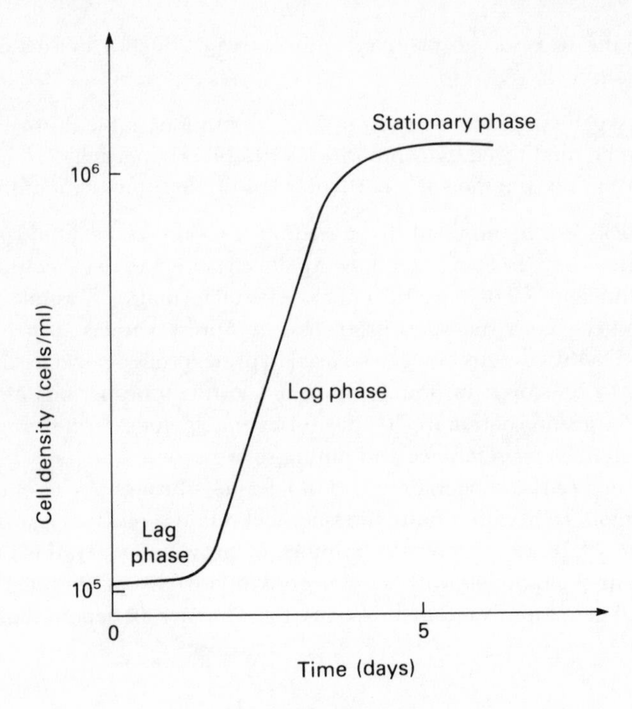

Fig. 1.4 Typical pattern of cell growth in culture.

In a typical batch culture, an initial cell inoculum of 10^5 cells/ml can continue growth up to 1 to 2×10^6 cells/ml, which involves a cell multiplicity (i.e. cell yield/cell inoculum) of 10–20. Cell concentrations *in vivo* can reach 10^9 cells/ml. However, these cells are in a homeostatic balance which is provided by the nutrient supply and inhibitor removal of the animal's vascular system. Attempts to increase cell yields *in vitro* can be made by mimicking this by perfusing medium through the culture, a system which so far has attained cell concentrations of $>10^7$ cells/ml.

The components required in cell culture media formulations are a complex mixture of carbohydrate, amino acids, salts, vitamins, hormones and growth factors (Table 1.1). Such formulations are far more complex than bacterial culture media where a single carbon source is often sufficient. Many of the formulations are based on modifications of that originally devised by Eagle, and most of these are adequate for small laboratory scale cultures. However, for scale-up to industrial production it is important to maximize cell yields. This involves a full understanding of the physico-chemical environment of the cell, a subject that requires a great deal more research so that optimal media formulations may be developed.

The media required for cell growth is an aqueous solution of nutrients and co-factors in a salt concentration which is isotonic. The constituent carbohydrate which is normally glucose but may be substituted by fructose or maltose, was once thought to be the sole energy source. However, evidence now shows that the amino acid carbon skeletons are also important as carbon energy sources — particularly glutamine. Cells require a continuous supply of amino acids and carbohydrate, even in the stationary phase. The complete consumption of any one of these media components may cause a nutrient limitation which may result in curtailment of growth or entry to a death phase. In most cultures, glutamine and glucose are utilized particularly rapidly and can cause growth limitation even before complete exhaustion.

Most of the glycolytic metabolism is anaerobic in cultured cells, and this leads to the accumulation of lactic acid in the medium. Buffering or some other pH control of the cultures is therefore essential to avoid drift to adverse acidic conditions. Bicarbonate is often included to act as a buffer system in conjunction with the gaseous carbon dioxide environment in which the cells can be cultured. Ammonia accumulation which results from deamination of the amino acids may cause a cell growth limitation because of its toxicity. The vitamin components of the medium are present in relatively low concentrations and are utilized as co-factors, the requirements for which show considerable variations between cell lines.

For most cultures, supplementation with blood serum (at 10%) is required to establish cell growth. The serum can be obtained from various animal sources — bovine or equine being the most common. Foetal calf serum is the most effective for cell growth promotion and is often used for cells whose growth is slow or difficult. Serum contains an uncharacterized mixture of ~1000 different proteins, many of which are factors that provide the required growth promoting properties. However, this requirement for serum has many disadvantages in cell cultures:

(1) It is expensive and may account for 70–80% of the overall media cost, particularly if foetal calf serum is used
(2) It is an undefined chemical mixture, the content of which can vary from batch to batch. This can give rise to inconsistent results between cultures
(3) It increases the overall protein content of the media which causes frothing of the culture if air is bubbled through. This can affect the viability of the cells
(4) The presence of a mixture of different proteins in the medium can cause problems in extraction of the desired cell-derived product during 'downstream processing'. These problems lead to the need for further purification steps which increase the overall cost of the process
(5) Serum is the major source of virus or mycoplasma contamination that can affect cultures. This type of contamination is particularly serious because it can lead to intracellular infections which result in product contamination and a gradual deterioration of the cell line

For these reasons there has been considerable work involved in attempts to formulate serum-free media. These formulations contain purified hormones and growth factors which can substitute for serum supplements. One example is the HITES medium, which contains hydrocortisone, insulin, transferrin, estrogen and selenite. Selected concentrations of these components in medium can substitute for serum in supporting the growth of some hybridoma cells (Chapter 6). However, each cell line is exacting in its requirements and formulations of this type may not have general application. Serum-free formulations have worked well for fast growing tumorigenic or transformed cell lines whereas the development of suitable serum-free formulations for growth of fastidious cells is more of a problem.

Summary

The historical development of cell culture techniques commenced with the classical experiments of Harrison in 1907. Further developments leading to the routine laboratory use of such techniques were aided by the use of trypsin for cell disaggregation, antibiotics to limit the problems of bacterial contamination and the formulation of semi-defined growth medium.

Cells which are not derived from biological fluids are anchorage-dependent and require a substratum for growth. Such cells taken directly from animal tissue form a primary culture which can be sub-cultured for a finite number of generations. Cell transformation, which is a process that can be induced by viruses or mutagens, leads to the loss of some of the growth restrictive characteristics of 'normal' diploid cells in culture. Of particular importance is the change to infinite growth capacity that allows such cells to be cultured indefinitely and may also allow the cells to grow in suspension. Cells can be stored in cryopreservatives at low temperatures and this allows the establishments of cell repositories which promotes the continued use of selected cell lines.

An inoculated culture will show cell growth through a lag phase to an exponential phase in which cells will continue growth until some limitation is encountered. This limitation may be growth surface area for anchorage-dependent cells, although nutrient depletion or inhibitor accumulation can lead to the stationary phase. A full understanding of the associated parameters, particularly media components, may lead to higher cell yields. The ideal media formulation should be entirely chemically defined, and thus avoid many of the disadvantages of supplementation with serum.

General Reading

Adams, R.L.P. (1980). 'Laboratory Techniques in Biochemistry and Molecular Biology — Cell Culture for Biochemists'. Barking, Elsevier Applied Science.

Barnes, D.W., Sirbasku, D.A. and Sato, G.H. (Eds) (1984). '*Cell Culture Methods for Molecular and Cell Biology*' Vols. 1–4, Alan Liss Inc.

Freshney, R.I. (1983). 'Culture of Animal Cells: manual of basic technique'. Alan Liss Inc.

Freshney, R.I. (Ed.) (1986). '*Animal Cell Culture: a practical approach*'. Oxford, IRL Press.

Harris, C.C., Trump, B.F. and Stoner, G.D. (1980). '*Normal Human Tissue and Cell Culture, Methods in Cell Biology*', Vol. 21. London, Academic Press.

Mather, J.P. (Ed.) (1984). 'Mammalian Cell Culture — the use of serum-free hormone-supplemented media'. New York, Plenum Press.
Paul, J. (1975). *Cell and Tissue Culture*. Edinburgh, Livingstone.
Sharp, J.A. (1977). *An Introduction to Animal Tissue Culture*. London, Edward Arnold.
Spier, R.E. and Griffiths, B. (Eds) (1985 and 1986). *Animal Cell Biotechnology*, Vols. 1 and 2. London, Academic Press.
Taub, M. (Ed.) (1985). *Tissue Culture of Epithelial Cells*. New York, Plenum Press.
Thilly, W.G. (Ed.) (1986). *Mammalian Cell Technology*, London, Butterworth.

Chapter 2

Biochemical Engineering Aspects of Culture Scale-Up

Cell Culture System Design — The Bioreactor

Non- Anchorage-Dependent cells

Animal cells differ from bacteria in the absence of a cell wall. This results in a greater degree of fragility which must be reflected in the way cells are treated in culture. However, animal cells which are non- anchorage-dependent may be grown in suspension in stirred tanks in a similar way to bacterial cultures. Suspension cells can be grown in small static non-stirred flasks but this offers only limited possibilities for scale-up.

Mass transfer of nutrients and oxygen requires efficient mixing of the culture so as to maintain homogenous conditions. A simple stirring procedure involves the rotation of a suspended bar by magnetic stirring in a spinner flask. This can be a suitable method of stirring up to 20 l beyond which motor driven stirrers are necessary. However, the shear force generated by the use of stirrers or impellers must be considered because of the potential damage to fragile animal cells. Large impellers rotating at low stirring speeds (typically upto 100 rpm) are preferable because of the low shear forces generated. As the scale of the culture is increased it becomes more important to ensure vertical mixing through the culture. This can be provided by an impeller which allows turbulence throughout the entire culture.

Many impeller designs are possible as shown in Fig. 2.1. One favoured design for animal cell cultures is the marine-type impeller which ensures the lift and radial mixing at the low stirring speeds required in these cultures. Adequate turbulence is also ensured by the use of round-bottomed vessels which are used in preference to the flat-based vessels used for bacterial cultures.

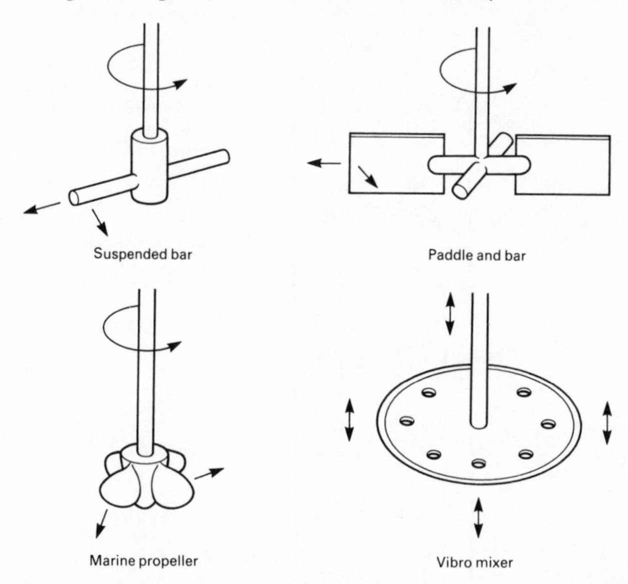

Fig. 2.1 Types of impellers that may be used in stirred tank reactors.

An alternative mixing system can be provided in a stirred tank by a Vibromixer. This produces a stirring effect by the fast reciprocating vertical motion of a mixing disc which has conical shaped holes (see Fig. 2.1). The shaft which holds the disc in the culture vibrates typically with a path of 2 mm at a frequency of 60 Hz (cycles/sec).

An alternative to the stirred tank bioreactor is the airlift fermenter which depends upon a stream of air at the bottom of a glass column to allow mixing and aeration. The column has a large height to diameter ratio. The inside of the column has a baffle or draught tube to allow the air flow to pass through the full length of the column (Fig. 2.2). A particular advantage of this culture system is that efficient stirring and aeration can occur without excessive shear forces. It is a simple system compared to the motor driven stirred tank bioreactors which often develop problems connected with the aseptic seal around the stirrer shaft at the point of entry into the fermenter. The airlift system has been used successfully for the growth of hybridoma cells on an industrial scale up to 1000 l by Celltech who have plans for further scale-up to 10 000 l (Chapter 6).

Anchorage-Dependent Cells

Anchorage-dependent cells pose an extra problem for culture, namely the surface substratum required for growth. The Petri dishes, Carrel flasks, T-flasks and Roux bottles used for small scale cultures are not amenable to scale-up and alternatives must be considered for industrial production processes. An important criterion for scale-up is the provision of a large growth surface to volume ratio. In the 1950s many industrial vaccine production plants were designed with a facility for cell culture using several thousand roller bottles. These are cylindrical containers which can be slowly rotated on mechanical rollers so that the inner surface

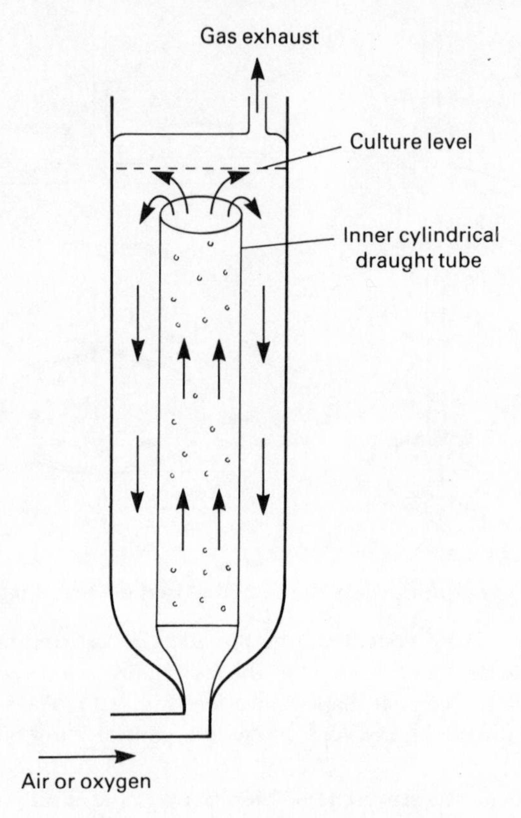

Fig. 2.2 The airlift fermenter.

may act as a cell surface substratum. The rotation ensures that the cells bound to the inner surface alternatively come into contact with the gas space and the growth medium in the bottles. The scale of operation in this system depends upon the number of roller bottles used, i.e. a multiple operation. The disadvantage of such a process is that it is labour intensive; this is expensive and the risks of contamination are enhanced because of the number of handling operations necessary.

Many alternatives can be considered to the roller bottle system. The most promising for industrial scale-up are those that involve a unit operation that can be increased in volume as scale-up is required. Table 2.1 shows examples of bioreactor systems that have been devised for anchorage-dependent cells. The corresponding growth surface to volume ratio is shown for each system. Of these examples, three have been found particularly suitable for scale-up as a unit operation — the artificial capillary, the glass bead packed bed and microcarrier culture systems (Fig. 2.3).

The Artificial Capillary System is based on the principle of pumping culture medium through bundles of synthetic hollow fibres to which the cells may attach. A large growth surface area is provided by the inner walls of the capillaries which may have diameters typically of 350 μm. A re-circulation system is operated so that the medium is continuously pumped through the capillary bundles. Such a system can

Table 2.1 Culture systems available for the growth of anchorage-dependent animal cells

Culture system	*Surface/volume ratio*
Roller bottle	1.25
Gyrogen	1.2
Multi-plate bottle	1.7
Spiral film bottle	4.0
Plastic bags	5.0
Tubular spiral film	9.4
Packed bed reactor	10.0
Artificial capillaries	30.7
Microcarriers (25 g l^{-1})	150.0

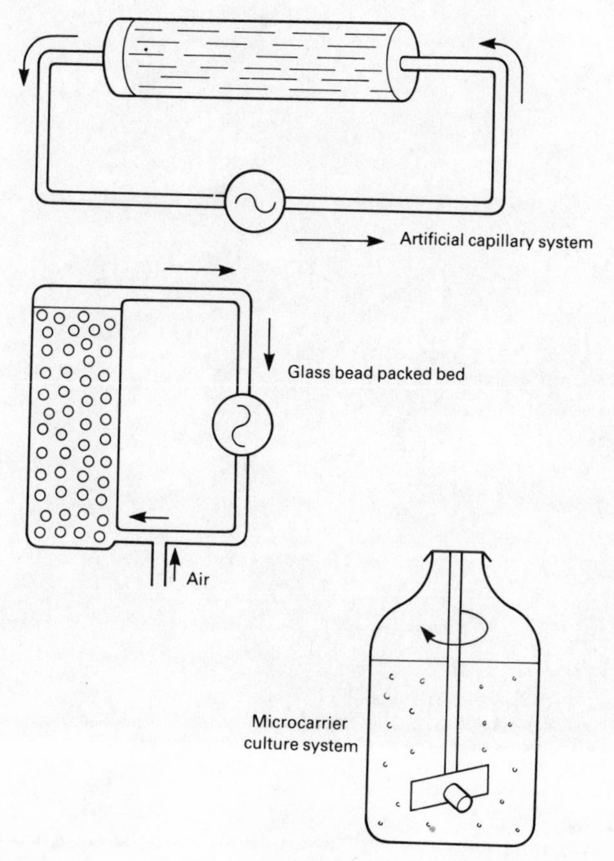

Fig. 2.3 Unit systems suitable for scale-up of anchorage-dependent cell cultures.

be provided from the capillary fibres made from an acrylic polymer as designed by Amicon originally for ultrafiltration. The fibre walls through which the medium is pumped provide a large surface area for cell attachment and growth. Alternatively, the Opticell system consists of a cylindrical ceramic catridge with 1 mm^2 square channels running lengthwise through the unit.

Capillary columns of this type have been designed upto a volume of 10 l. However, scale up for industrial production may be limited by difficulties of cell growth determination and equipment sterilisation.

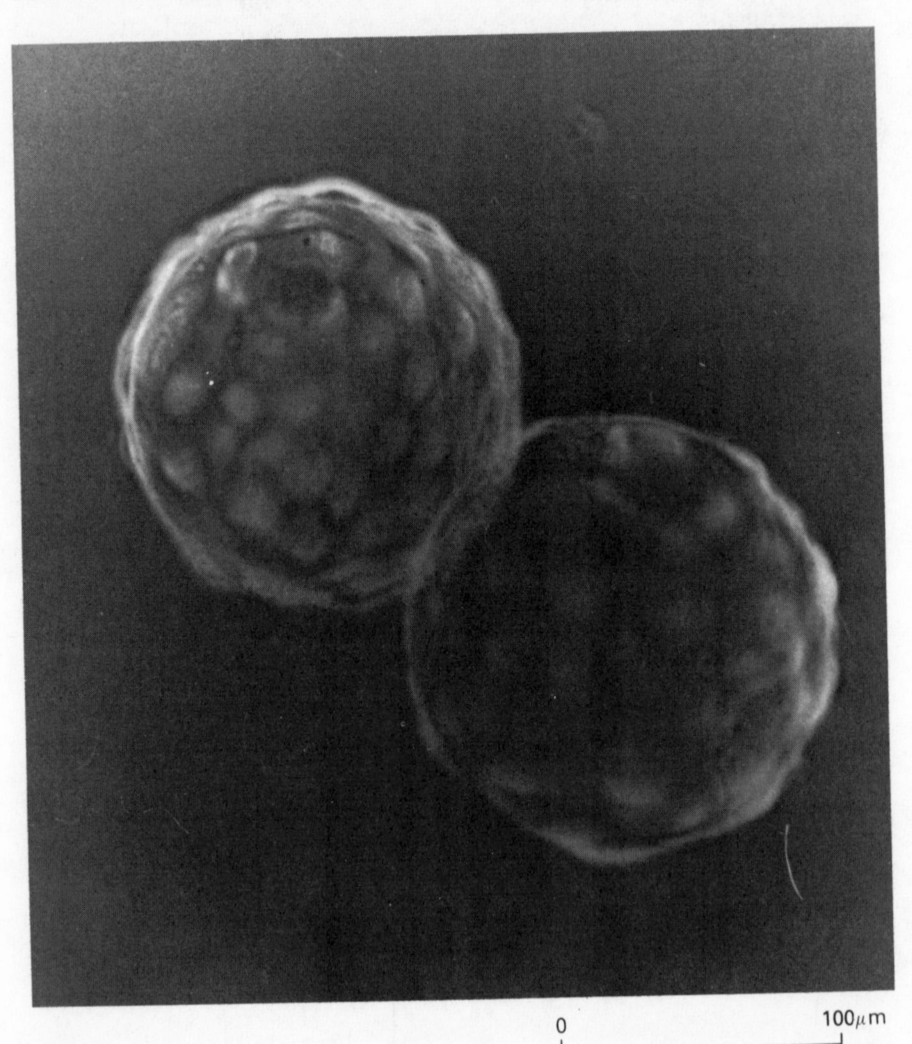

Fig. 2.4 Two microcarriers with a confluent monolayer of canine epithelial cells.

The Glass Bead System involves the use of 3–5 mm diameter glass beads packed into a glass column through which media can be re-circulated. Aeration and monitoring of the medium can be conducted outside the column. On inoculation, the cells attach to the exposed glass bead surface which provides a suitable substratum for cell attachment and growth. Such a packed bed bioreactor has been operated successfully at a 100 l scale for the culture of BHK cells for the preparation of Foot-and-Mouth disease virus. However, the major disadvantages of this system are that:

(1) Cells can not be sampled directly and so growth must be measured indirectly, e.g. by glucose consumption
(2) It can be difficult to maintain an even distribution of cells throughout the packed bed column. This problem may increase with scale up

Microcarriers are spherical beads of 150–200 μm diameter which can be suspended in a culture medium. Cells can attach and grow on the bead surface as a monolayer of up to 100–200 cells per microcarrier (Fig. 2.4). This system of culturing anchorage-dependent cells is particularly attractive because of the large surface to volume ratio available. Also, the cultures can be contained in the typical stirred tank bioreactors used for suspension cells. The microcarrier culture system is referred to as quasi-suspension (the well-stirred culture being homogeneous), and this has the advantage of allowing easy sampling for cell counts or any other culture parameter. The original microcarriers used were DEAE-Sephadex particles commonly used for ion exchange chromatography. However, the surface charge density of these was too high for cell attachment and resulted in an apparent cytotoxicity. Lower surface charge density dextran beads have now been used extensively, e.g. Cytodex. Other materials have also been found to be suitable. These include solid polystyrene, cellulose, gelatin and glass-coated plastic. The latter are useful because of their re-usability and ease of cell removal.

Microcarrier cultures are generally considered to have the greatest advantages for the scale up of anchorage-dependent cells. Simple stirred tank batch fermenters can be easily adapted to suit the mixing conditions required to maintain a homogeneous culture of microcarriers, and to date such cultures have been operated on a scale up to 1000 l. The cell yield from such batch cultures typically reaches a maximum of $\sim 10^6$ cells/ml. However, at these cell densities, the surface area offered by a concentration of microcarriers >5 mg/ml is not completely utilized. Medium re-feeding or perfusion can raise cell yields to $>10^7$ cells/ml. Thus it is important to appreciate that it is not only the volume of the bioreactor that determines the scale of operation — a 100 l microcarrier culture operated by medium perfusion may produce a higher yield than a 1000 l culture operated in a batch mode.

Despite the greater complexity of culture systems designed for growing anchorage-dependent cells there can be some advantages of using such cultures compared to suspension cultures. Media changes and perfusion are much easier because there is less of a problem of cell wash-out. Such media changes may be required to induce cells to secrete products. Also, many cells express products more readily when they are attached to a substratum.

Oxygen

The supply of oxygen to the cells in culture is a particular problem when culture volumes are increased from the laboratory scale. The solubility of oxygen in aqueous solution is relatively low — 0.2 mmol per litre of culture. Thus at a typical consumption of 0.05–0.5 μmol oxygen/10^6 cells/h, a one litre culture at 10^6 cells/ml would be depleted of oxygen in 0.5–4 h unless a constant oxygen supply is provided.

In small laboratory scale cultures the surface area to volume ratio of the culture is sufficiently high to allow oxygen to diffuse through the liquid surface from the gas of the head space of the flask. The rate of oxygen supply by such diffusion is governed by the oxygen transfer rate (OTR) which can be determined from equation (2):

$$\text{OTR} = K_L A(C^* - C_L) \qquad (2)$$

where K_L = mass transfer coefficient (cm/h)
A = interfacial area per unit volume (cm^{-1})
C^* = oxygen concentration of the solution at equilibrium with the gas
= phase (mmol/l)
C_L = the prevailing oxygen concentration in culture

Thus the rate of oxygen transfer is proportional to the concentration difference of oxygen ($C^* - C_L$) between the actual and equilibrium values. Also, the rate of mass transfer of oxygen is directly proportional to the upper surface area of the culture.

At the slow agitation rates normally used for mammalian cell cultures, the mass transfer coefficient is low. Figures in Table 2.2 show that at a constant depth to surface diameter ratio (aspect ratio), the culture surface area to volume ratio decreases significantly with increasing culture volume. Calculations show that above a critical volume of around 1 litre, oxygen diffusion via the head space becomes insufficient to satisfy a culture at a cell concentration of 10^6 cells/ml. There are several alternative approaches that may be considered for oxygen supply to the culture.

Oxygen In The Head Space. By filling the head space of the culture with 100% oxygen the rate of diffusion will increase by a factor of × 5. Although this does improve oxygen supply to the cells, this approach is expensive and can not be used for indefinite scale-up.

Direct Aeration into the culture by a sparger is the method of oxygen supply used in most bacterial cultures. However, in animal cell cultures this method may cause excessive foaming which may result in damage to the cells. This is particularly a problem with microcarriers which become trapped in the foam layer. As it is the protein content of the medium that is responsible for such foaming, the problem may be overcome by using serum-free medium. Foam formation may also be reduced by the addition of anti-foam reagents to the medium, e.g. Pluronic (polyglycol). An alternative is to reduce the bubble formation as oxygen is supplied to the culture. This has been shown to be successful in a system which has an air sparger surrounded by a fine mesh cage, or in the use of fine pore silicone tubing that can be suspended in culture.

Indirect Aeration. In culture systems that require medium re-circulation to allow adequate mixing, e.g. the glass bead or artificial capillary systems, aeration may be performed in a secondary vessel or even in tubing away from the fermenter. Such an external oxygenator to the culture can alleviate many of the problems of foam damage to the cells.

Table 2.2 The limitation of oxygen supply by diffusion through the head space

Culture volume (l)	*Head space area (cm^2)*	*Oxygen supply by from head space ($mMol\ h^{-1}$)*	*Cellular O_2 demand ($mMol\ h^{-1}$)*
1	100	0.063	0.063
10	500	0.313	0.625
100	2500	1.56	7.81

Notes
These calculations are based on the following assumptions:
A culture with liquid height (H)/diameter (D) = 1
Cell concentration in culture = 10^6 cells per ml,
and the following experimentally determined values:
Head space aeration = 0.63 μMol O_2 cm^{-2} h^{-1}
Oxygen utilization rate = 0.063 mMol O_2 l^{-1} h^{-1}.
Source (from Katinger and Scheirer, 1985)

Continuous Cultures

The cultures described so far can be divided into two categories — homogeneous or heterogeneous cultures. The homogeneous cultures include the non-anchorage cell suspension and microcarrier cultures. The efficient mixing that can be assured in such cultures allows simple sampling techniques and a fast distribution of any substrate or factor that is added. The heterogeneous culture systems include the glass bead and artificial capillary systems. In such heterogeneous systems, nutrients are gradually depleted as they flow through the anchored cells and this may give rise to a concentration gradient — a phenomenon typical of a 'plug-flow' system. To minimize the possibility of developing such gradients, the medium is usually re-circulated (Fig. 2.3).

Batch cultures and the re-circulating systems described follow kinetic patterns as shown in Fig. 2.5. The nutrient concentrations of the medium follow an exponential decrease which mirrors the accumulation of metabolic products. Cell growth will continue in such culture systems until the depletion of a nutrient or the accumulation of a growth inhibitory product reaches a critical concentration that arrests further cell division. In order to increase the supply of nutrients and removal of growth inhibitors a perfusion system may be used. In such a system the cells are entrapped in the fermenter whilst medium is continuously pumped through the culture. This has the effect of stabilizing the concentration of nutrients

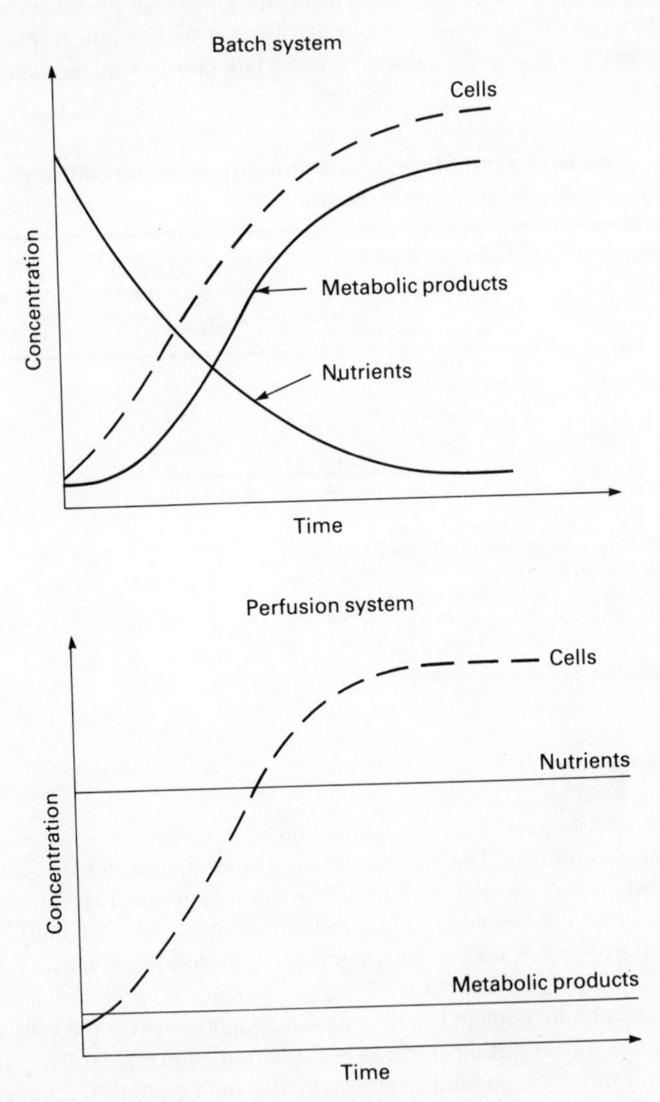

Fig. 2.5 Kinetics of nutrient consumption and product accumulation in batch and perfusion cultures.

and removing inhibitory metabolites. This can result in a higher cell yield which is dependent upon the velocity of the medium flow per unit volume and described as the dilution rate:

$$\text{Dilution rate} = F/V$$

where F = medium flow rate (l/h)
V = volume of culture (l).

In a heterogeneous system the cells are already entrapped on a solid matrix. In homogeneous systems some means of cell containment is required. This may take the form of a spin filter which is a rotating mesh through which media may be pumped but which limits cell removal. In the case of microcarrier cultures media may be pumped from a settling column in which the microcarriers are allowed to sediment back into the culture. A typical system for this is shown in Fig. 2.6.

The kinetics of such a continuous culture was analysed by Monod for bacterial cultures. He suggested that the rate of cell growth may be dependent upon the rate of supply of a critical substrate required for growth. The equation applied to this is analogous to that devised by Michaelis-Menten for enzyme kinetics:

$$\lambda = \lambda\text{max} \cdot S / (S + K\text{s}) \qquad (3)$$

where λ = specific cell growth rate = $(\mathrm{d}x/\mathrm{d}t) \cdot 1/x$, i.e. rate of cell mass increase $(\mathrm{d}x/\mathrm{d}t)$ per unit of cell mass (x).
λmax = maximum specific cell growth rate
Ks = Michaelis-Menten constant
S = concentration of limiting substrate.

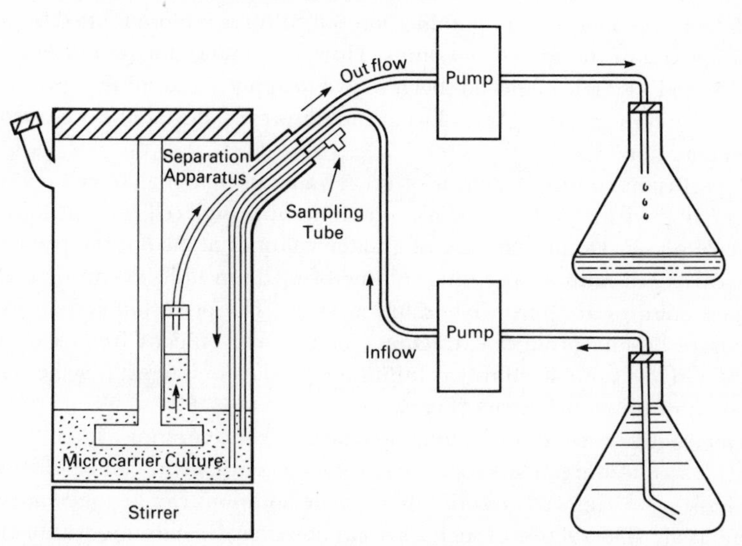

Fig. 2.6 A medium perfusion culture system suitable for cell growth on microcarriers.
(From Butler, 1983)

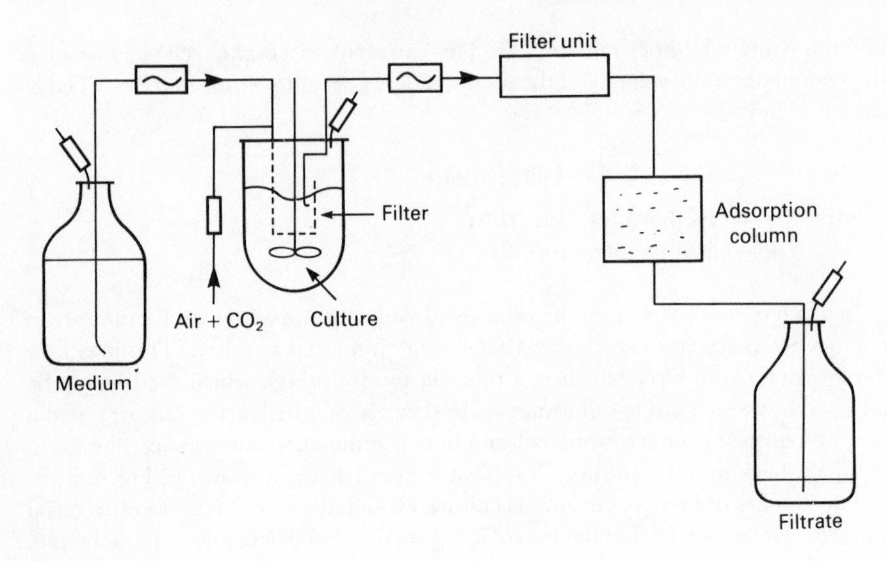

Fig. 2.7 In line product extraction.
(From Kluft, 1983)

This kinetic model assumes that cell growth is affected by the concentration of a substrate, *S*. If the concentration of *S* is above a critical value then cell growth will proceed at a maximum rate. The growth rate decreases with the decrease in the concentration of *S*. In a perfusion culture the concentration of *S* is dependent upon the dilution rate.

This model has been applied successfully to continuous growth in bacterial cultures where a single carbon substrate can usually be identified as the limiting growth factor. Application to mammalian cell cultures is more limited because of the complexity of the growth medium. However under glucose limiting conditions, Monod's kinetic model has been shown to apply to cell cultures over a range of low dilution rates. At higher dilution rates however, deviations occur from the Monod equation.

The perfusion culture system is of great value in mammalian cell cultures in attaining high yields ($>10^7$ cells/ml) or maintaining high cell concentrations in a stationary phase. Despite the use of greater volumes of media, the productivity expressed as cell number per unit volume of medium utilized can be higher in perfusion cultures at appropriate dilution rates. The perfusion system can also incorporate on-line product extraction. Thus if the effluent from a fermenter is connected to a suitable affinity column, the product of interest may be removed from the flow of 'spent' media (Fig. 2.7).

An alternative system for stabilizing nutrient concentrations in culture is by selective batch feeding. Thus if particular media components can be identified as being limiting during cell growth, appropriate additions can be made at selected time intervals. The success of such a system however depends upon a full analysis of the nutrient requirements of a particular cell during growth.

Computer Control of Culture Parameters

During cell growth in any of the culture systems discussed in this chapter, there are a range of parameters which are required to be kept at their optimal value to ensure maximum cell growth. Some of these parameters can be measured directly by the use of sterilisable probes inserted into the fermenter. These directly measurable parameters include temperature, pH, oxygen and foam level. In such cases, the amplified electrical reading from each probe can be transmitted electronically to a control unit which can compare the reading with a pre-determined set-point. Deviations of the probe output from the set-point will activate some form of corrective measure to shift the culture parameter towards the optimal conditions. This can involve the activation of a pump to feed alkali, anti-foam or oxygen into the culture or switching on a heating system.

Figure 2.8 shows how such a feedback control system can be operated to maintain optimum pH (usually 7.4). Deviations from this value will be recognised by the control unit from the readings of the pH probe. Corrective measures may be

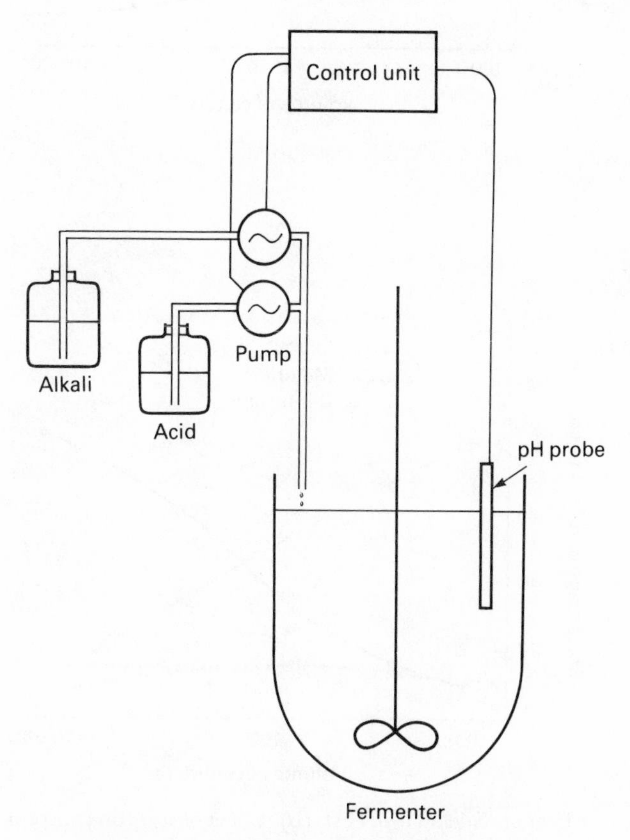

Fig. 2.8 Feedback system for pH control.

taken by activating a pump which will allow pre-determined doses of acid or alkali into the culture until the optimum pH value is restored. Similarly, oxygen and foam levels may be controlled by feedback systems which pump air or inject anti-foam into the culture. The optimal temperature is normally maintained by the external air temperature or by circulating water around an outer jacket of the

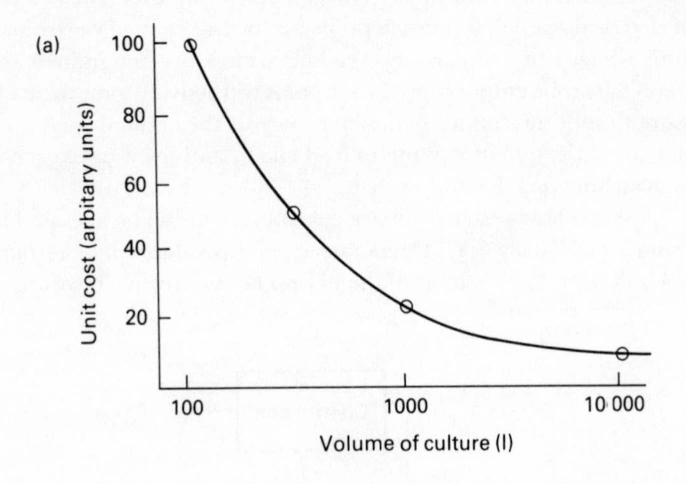

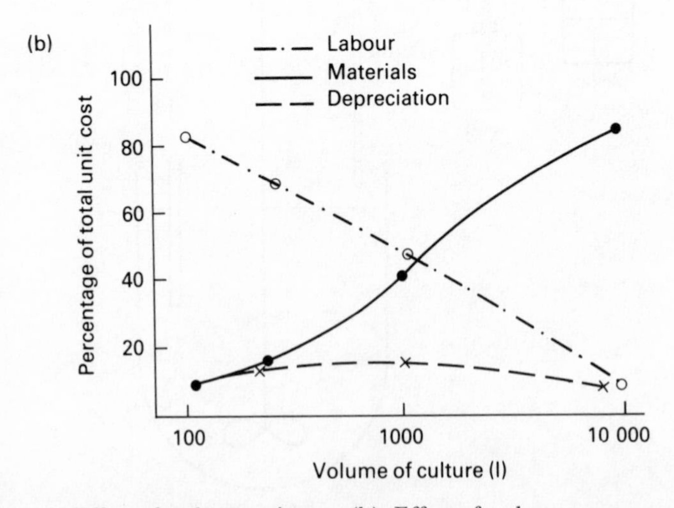

Fig. 2.9 (a) Effect of scale on unit cost. (b) Effect of scale on components of unit cost.
(From Birch, *et al.*, 1987)

fermenter. A thermal detector in the culture can be linked to an external heating unit so as to ensure the maintenance of the optimal temperature of the culture.

There are a number of computerized control units of this type now commercially available. Such units will continuously monitor the appropriate parameters and take necessary corrective action as pre-determined by the operator before a culture run. The limitation to further developments of such control is the availability of suitable probes for continuously measuring culture parameters. It would undoubtedly be advantageous to control a range of critical factors such as nutrient levels by controlled feeding from pumps. This would enable continuous nutrient re-feeding so that critical nutrients such as glucose and amino acids would be maintained at their optimal concentrations in culture. Such an optimal environment for cell growth and metabolism could also limit the accumulation of those adverse metabolic products such as lactic acid and ammonia.

Cost Analysis of Scale-Up of Cell Cultures

All commercial operations require detailed cost analysis: an aspect not always considered at the laboratory scale. Cell culture is labour-intensive and this is normally reflected in the cost of a large-scale operation. In a multiple process such as the roller bottle culture system, the labour demands increase directly with scale-up. Thus, the number of handling operations required in establishing 1000 roller bottle cultures is × 10 that of 100 cultures. However, for a unit process, the labour required does not increase proportionally with scale — the demands of managing a 1000 l fermenter are not × 10 that of a 100 l. This is reflected in the unit cost (cost per gram of product) which decreases with scale-up. For this reason a unit process is favoured over a multiple process for large-scale production.

With the scale-up of a unit process, material costs become increasingly significant as a proportion of the overall operation. Figure 2.9 shows data made available by Celltech on the relative costs of scale-up of an airlift fermenter system. This shows that the depreciation of capital cost of the equipment does not vary with scale-up. However, a significant change occurs at the 10^3 l scale above which material costs exceed labour costs as the single most expensive component of operation.

These material costs relate particularly to the culture media which is typically between × 5 and × 20 more expensive for animal cell compared to bacterial cell cultures. A detailed analysis of the media costs for animal cell cultures is given in Table 2.3 which shows a breakdown of costs of media components expressed as a percentage of the overall cost. These figures show that the antibiotic and serum components of the media can account for up to 97% of the total.

In order to reduce media costs, several approaches may be considered. Omission of antibiotics may be a possibility and even desirable. This should be acceptable with the use of careful procedures for culture manipulation — which in any case should be to a standard to prevent antibiotic resistant contaminants, e.g. viruses and mycoplasmas. Consideration could be given to the type and concentra-

Table 2.3 Percentage cost of constituents of growth medium

	Foetal calf serum	*Calf serum*
Basic medium	3	12
Serum	84	44
Antibiotics	13	44
Total	100	100

(From Griffiths, 1986)

tion of serum required. The figures in Table 2.3 are based on media with a typical 10% serum supplement of either foetal calf or calf serum. The use of a lowered serum content or an alternative source (particularly other than foetal calf serum) would enable significant cost reductions.

Serum-free media formulations may be suitable for the growth of certain cell lines. However, these formulations may contain certain hormones and growth factors, the combined cost of which may exceed that of the serum supplement. However, the use of such a serum-free formulation may still be justified in a large-scale operation if there are benefits to the later stages of product purification (see Chapter 1).

In assessing the commercial costs of a cell culture operation it is important to consider all possible strategies. The important parameter to an industrial enterprise must be product generated per unit cost of operation. Any means of increasing this productivity value would be desirable.

In many cases it may be possible to increase such productivity by consideration of cell selection or cell culture strategy. These may indeed be alternatives to the culture volume scale-up often considered as the only option. In the first instance, careful selection of a cell line for high protein production or amplification of an appropriate gene may make productivity differences of × 10 to × 100 (see Chapter 3).

The choice of media perfusion rather than batch culture can lead to significant increases in cell yields. Though perfusion may require the expenditure of larger volumes of media, it may be possible to use diluted media or carefully controlled dilution rates to improve overall productivity. The relationship between cell yields and perfusion rates has shown that the supply of nutrients and the removal of toxic products from these cultures can increase cell productivity by an order of magnitude.

The positive identification of those factors that allow such yield increases must be the key to allow further major increases in productivity in cell cultures. By the formulation of optimal growth media or the development of a regime of selected nutrient addition and inhibitor removal, there would seem to be no logical reason why higher yields could not be routinely obtained. Considerations of this sort in planning culture operations may significantly alter the bioreactor volumes necessary and reduce the overall unit cost of culture.

Summary

Some animal cells — notably transformed cells and cells derived from biological fluids — can be grown in suspension. Simple stirred tank reactors or airlift fermenters are suitable for such cultures which can be scaled up easily. The scale-up of anchorage-dependent cells is a little more problematic and involves consideration of the substratum for cell growth. The most promising systems for the large-scale growth of these cells have been unit reactors such as packed beds, hollow fibres or microcarrier cultures.

At the laboratory scale, oxygen diffusion through the culture surface is adequate to supply the growth needs of cells. However, above the 1 litre scale some other form of oxygen supply is required. There are various solutions; the most promising of these include direct oxygen or air supply via a fine mesh or through capillary tubing. These systems avoid the problem of frothing in culture which may cause cell damage.

The use of continuous perfusion can increase the final cell yield by the supply of nutrients and removal of inhibitors from the culture. This yield is dependent upon the dilution rate which if optimized can increase the productivity of a system in terms of cells produced per unit volume of utilised media.

The use of sterilisable probes in fermenters enables the continuous monitoring of culture parameters such as pH, temperature, oxygen and foam. With computer controls, such information can be used to initiate corrective action to maintain these parameters at their optimum values during cell growth.

Detailed analysis of a large-scale culture operation can suggest strategies to reduce the overall unit cost. In particular, considerations of appropriate cell selection, scale of culture and need for serum supplementation could make extensive differences to product costs.

General Reading

Arathoon, W.R. and Birch, J.R. (1986). 'Large-scale cell culture in biotechnology', *Science* **232**, pp. 1390–1395.

Feder, J.B. and Tolbert, W.R. (1983). 'The large-scale cultivation of mammalian cells', *Sci. Amer.* **248**, pp. 24–31.

Feder, J.B. and Tolbert, W.R. (Eds) (1985). *Large-scale Mammalian Cell Culture*. London, Academic Press.

Glacken, M.W., Fleischaker, R.J. and Sinskey, A.J. (1983). 'Mammalian cell culture: engineering principles and scale-up', *Trends in Biotech.* **1**, pp. 102–108.

Mizrahi, A. (1986). 'Biologicals from animal cells in culture', *Biotechnology* **4**, pp. 123–127.

Spier, R.E. (1980). 'Recent developments in the large scale cultivation of animal cells in monolayers', *Adv. Biochem. Eng.* **14**, pp. 119–162.

Spier, R.E. and Griffiths, B. (Eds) (1985 and 1986). *Animal Cell Biotechnology*, Vols. 1 and 2. London Academic Press.

Thilly, W.G. (Ed.) (1986). *Mammalian Cell Technology*. London, Butterworth.

Specific Reading

Buckland, B.C. (1984). 'The translation of scale in fermentation processes: The impact of computer process control', *Biotechnol.* **2**, pp. 875–882.

Griffiths, B. (1986). 'Call cell culture media costs be reduced? Strategies and possibilities', *Trends in Biotech.* **4**, pp. 268–272.

Knakek, R.A. *et al.* (1972). 'Cell culture on artificial capillaries: An approach to tissue growth *in vitro*', *Science* **175**, pp. 65–67.

Nardelli, L. and Panina, G.F. (1976). '10 years experience with a 28,000 roller bottle plant for FMD vaccine production', *Develop. Biol. Stand.* **37**, pp. 133–138.

Reuveny, S. (1985). 'Microcarriers in cell culture: Structure and application', *Adv. Cell Culture* **4**, 213–247.

Whiteside, J.P. *et al.* (1979). 'Development of a methodology for the production of Foot and Mouth Disease virus from BHK21 C13 monolayer cell growth in a 100 l ($20m^2$) glass spherical propagator', *Develop. Biol. Stand.* **42**, pp. 113–120.

Chapter 3

Genetically Engineered or Modified Cells

Recombinant DNA Technology

Many of the important techniques for the genetic manipulation of cells by recombinant DNA technology were developed in bacteria — particularly *E. coli*. These techniques have enabled the insertion of mammalian genes into bacterial cells and this has allowed the large-scale production of mammalian proteins in bacterial cultures. The first of these to be licensed for large-scale human consumption was insulin in 1982. Recently, these recombinant DNA techniques have been applied to mammalian cells and there are good reasons to suggest that such genetically engineered mammalian cells will be increasingly used for large-scale production processes. In this chapter the basic principles of recombinant DNA technology will be described. Later chapters discuss the application of this technology to the production of specific animal cell protein products.

The basis of recombinant DNA technology was established in the 1970s with the discovery and isolation of several useful enzyme types which enable the *in vitro* manipulation of DNA fragments. These key enzymes enable DNA to be cut, joined or polymerized and include:

(1) Restriction endonucleases (type II)
(2) DNA ligases
(3) RNA-directed DNA polymerase (reverse transcriptase)

DNA Restriction

This involves the fragmentation of DNA by one of a family of enzymes called res-

Enzyme	Recognition site	Micro-organism of origin
Eco RI	–G↓A–A–T–T–C– –C–T–T–A–A↑G–	*Escherichia coli*
Hae III	–G–G↓C–C– –C–C↑G–G–	*Haemophilus aegyptius*
Hha I	–G–C–G↓C– –C↑G–C–G–	*Haemophilus haemolyticus*
Pst I	–C–T–G–C–A↓G– –G↑A–C–G–T–C–	*Providencia stuartii*
Pvu II	–C–A–G↓C–T–G– –G–T–C↑G–A–C–	*Proteus vulgaris*
Sal I	–G↓T–C–G–A–C– –C–A–G–C–T↑G–	*Streptomyces albus*
Sau 3A	↓G–A–T–C– –C–T–A–G↑	*Staphylococcus aureus*
Xmn I	–G–A–A–N–N↓N–N–T–T–C– –C–T–T–N–N↑N–N–A–A–G–	*Xanthomonas manihotis*

↓ indicates cleavage N = any nucleotide

Fig. 3.1 Recognition sites and points of cleavage of restriction endonucleases.

triction endonucleases which are derived from micro-organisms. Of particular value for *in vitro* use are the type II restriction enzymes because of their ability to cause DNA cleavage at specific nucleotide sequences called recognition sites. Several hundred different types of these endonucleases have been isolated in a pure form and are available commercially.

Figure 3.1 depicts the activity of several of these enzymes which are typical of those that can be used for controlled DNA fragmentation. Each enzyme is derived from and named after one particular microbial species. The recognition site in double-stranded DNA is a sequence of nucleotides which can be of various lengths — hexanucleotides being quite common. Of the endonucleases listed in Fig. 3.1, recognition sequences are shown to vary from 4 base pairs (HaeIII) to 10 base pairs (XmnI). The fragmentation can be staggered and result in short single strands of DNA at each end of the break — known as cohesive (or sticky) ends, e.g. EcoRI and PstI. Alternatively, fragmentation can lead to a clean break across the double strand leading to blunt ends as as with PvuII, HaeIII and XmnI.

The length of the recognition sequence determines the average DNA fragment size which can be generated. Thus, assuming a random sequence distribution of nucleotides in DNA, a statistical prediction can be made of the frequency of occurrence of a specific hexanucleotide such as those shown in Fig. 3.1. With 4 different nucleotides in DNA, the frequency of occurrence of a specific hexanucleotide sequence is 4^6, i.e. 1 in 4096 nucleotides. Therefore, 4 kilobases will be the mean size of DNA fragments after treatment with an enzyme having such a recognition site. Thus for a mammalian cell with a DNA content of 3×10^9 base pairs, treatment with such an enzyme should result in 7.5×10^5 unique fragments. This assumes that the fragmentation reaction runs to completion. Partial digestion by an endonuclease is also useful and results in a wider distribution of fragment sizes.

DNA Ligation

The single-stranded sticky ends are useful in that two DNA fragments from different sources but cleaved by the same enzyme will have complementary sticky ends which can be induced to hybridize. Such hybridization will occur spontaneously on mixing the two fragments — a process called annealing. Once annealed, a covalent phosphodiester linkage can be formed between the two fragments by the enzyme (DNA ligase). Such enzymes can be obtained from microbial extracts, the

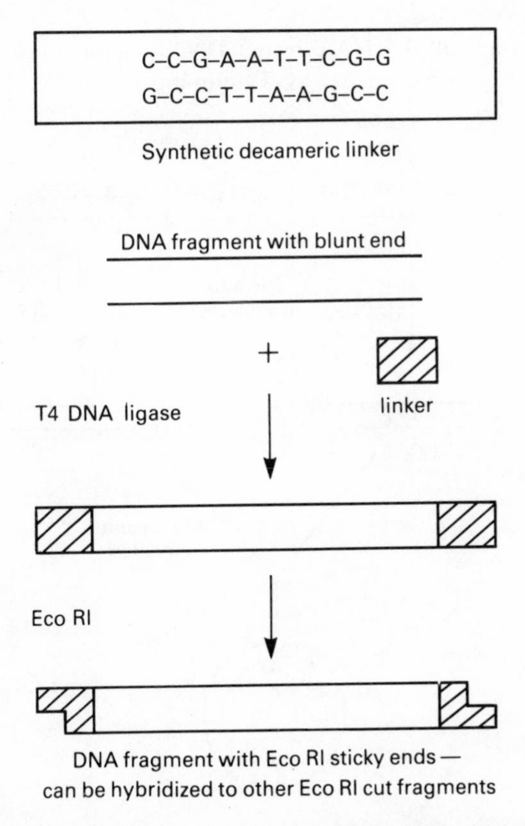

Fig. 3.2 Linkers to join blunt end fragments of DNA.

two most commonly used being derived from T4 bacteriophage and *E. coli*. These enzymes are available commercially in a pure form.

Sometimes it is necessary to link fragments without sticky ends. This can arise with the use of endonucleases that produce blunt end fragments or in the case of DNA fragments derived from two different endonucleases. In such cases, linker molecules can be added to the DNA terminals (Fig. 3.2). These may contain hexanucleotide recognition sequences related to the single strands normally produced as sticky ends. Once two sets of fragments have the same sticky ends they can be annealed and then firmly linked by the ligase enzymes.

Alternatively, homopolymer tailing is a useful procedure whereby a short sequence of a single nucleotide (e.g. -G-G-G-G-G-) is added to one set of fragments and a sequence of complementary nucleotides (e.g. -C-C-C-C-C-) added to the second set. This requires the enzyme — (terminal transferase) — to add nucleotides to the 3′ ends of the DNA fragments. Mixing of the two sets of fragments will cause hybridization of the complementary homopolymer tails which can then be treated with a ligase enzyme. This leads to a stable set of recombinant DNA fragments (Fig. 3.3).

Reverse Transcription

In 1970 an enzyme called RNA-directed DNA polymerase (or reverse transcriptase) was isolated from viruses by Temin and Baltimore. The value of this

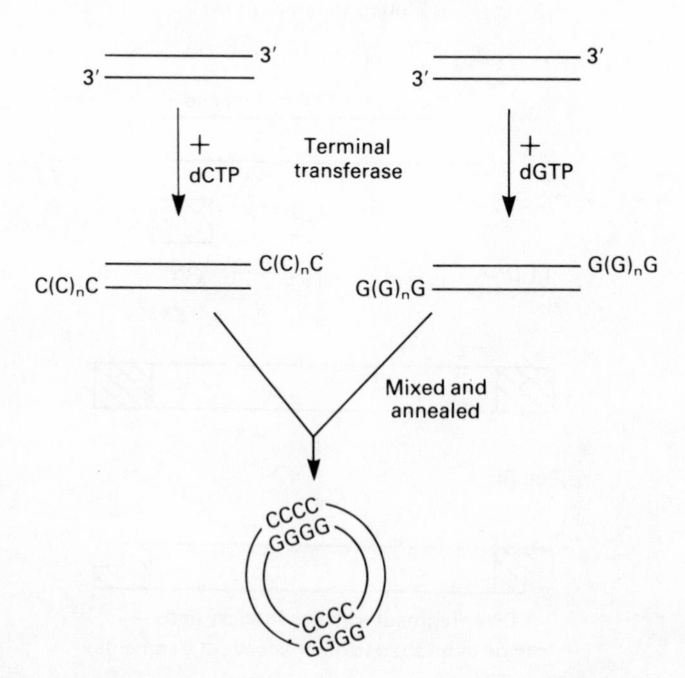

Fig. 3.3 Homopolymer tailing to join two independent strands of DNA.

isolated enzyme in DNA manipulation is its ability to form DNA from an RNA template. This enables complementary DNA (cDNA) to be formed from a cellular extract of mRNA. The stages involved in this process are shown in Fig. 3.4 and are as follows:

(1) mRNA isolated from mammalian cells has a polyadenylate tail — a sequence of 10–20 adenosine nucleotides at the 3′ end. A similar sized synthetic oligo-dT sequence is hybridized to this tail. This acts as an initiator for subsequent polymerization.
(2) In the presence of the 4 deoxyribonucleotides, reverse transcriptase will extend the oligo-dT strand by forming a sequence of nucleotides which are complementary to the mRNA

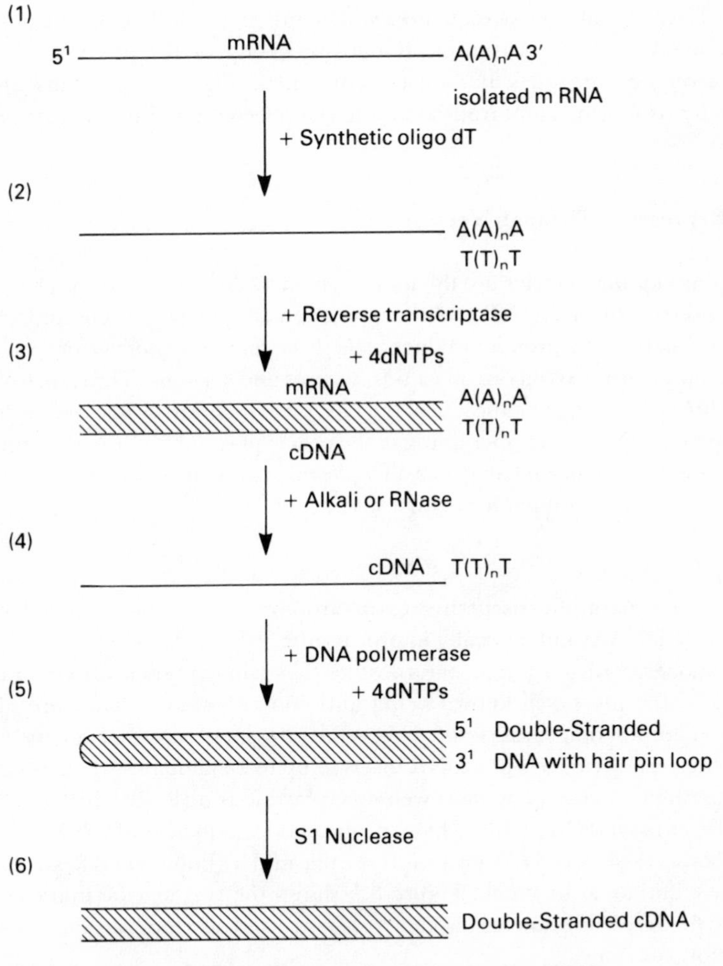

Fig. 3.4 Formation of cDNA from mRNA.

(3) The mRNA is selectively degraded from the double-stranded mRNA/DNA hybrid by alkali or RNAase treatment.
(4) The single-stranded cDNA is converted into the double-stranded form by incubation with DNA polymerase in the presence of the 4 deoxyribonucleotides. The mechanism of this is that a hairpin loop of a continuous nucleotide sequence is formed at the 3′ end
(5) The hairpin loop is finally broken by the action of a specific nuclease (S1) which results in the formation of the double-stranded complementary DNA

This double-stranded cDNA reflects the population of mRNA of the cell from which extraction was made. For example, cDNA produced from reticulocyte mRNA will contain predominantly sequences related to the globin gene. However, such a sequence will not be an exact transcript of the gene as it exists in the nuclear DNA — all control sequences will be missing as will be the introns which are removed in the post-transcriptional processing of the mRNA. However, cDNA sequences are particularly useful for genetic engineering because they can be directly transcribed and translated into the selected protein in a host cell.

Gene Expression Through Vectors

Vectors or cloning vehicles are the names given to those elements of DNA which can be inserted into a host cell to allow expression of a particular gene and its subsequent translation to a protein product. The three main types of vectors used in the genetic engineering of cells are plasmids, viruses and cosmids. The choice of which type to use is dependent on the size of the DNA fragments to be inserted — the size of acceptable DNA inserts increasing in the order: plasmid < virus < cosmid. For the expression of mammalian genes only plasmids and viruses have been used and discussion will be confined to them.

Procaryotic Cell Vectors
Plasmids are naturally occurring extrachromosomal double-stranded circular elements of DNA found naturally in procaryotic cells and varying in size from 2.8 to 900 kilobases long. In their natural state they can express a variety of characteristics — the most well known being antibiotic resistance. They are ideal for genetic manipulation because of their relatively small size and ability for self-replication. They can accept a DNA insert of up to 15 kilobases which is sufficient for a mammalian gene. The most well used plasmid is pBR322 which is a double-stranded circular DNA with a known nucleotide sequence of 4362 base pairs. pBR322 was re-structured from a wild type plasmid by Bolivar and Rodriguez who lent their initials to its name. Figure 3.5 shows the two genetic markers in this plasmid for ampicillin and tetracycline resistance and also four recognition sites for restriction enzymes.

Figure 3.6 shows the stages involved in the insertion of a mammalian gene into

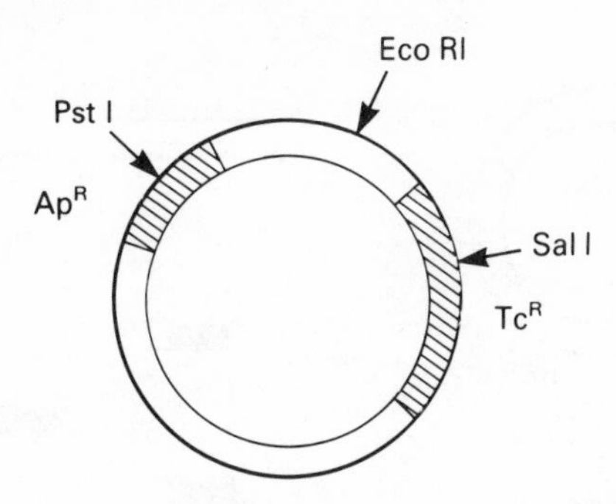

Fig. 3.5 Structure of the circular double-stranded DNA of the plasmid, pBR322.

the plasmid pBR322. Treatment of the plasmid with a restriction enzyme can convert the plasmid into a linear double-stranded length of DNA. This can then be combined with a foreign fragment of DNA by use of cohesive ends or by homopolymer tailing. One useful site for cleavage is the PstI recognition site, because of its position in the ampicillin resistant gene (Ap^R). The source of a mammalian gene suitable for expression is usually a cDNA transcript from mRNA derived from cells secreting the protein of interest. Insertion of such a cDNA transcript into the PstI site of pBR322 inactivates the Ap^R gene so that the recombinant plasmid retains its resistance to tetracycline but becomes ampicillin sensitive. This process of 'insertional inactivation' is a useful genetic marker for the recombinant plasmid (Fig. 3.6).

Most bacteria such as *E. coli* can take up DNA molecules from the medium. This process is known as transformation in procaryotic cells and should not be confused with the alternative meaning of transformation as used for eucaryotic cells (Chapter 1). The efficiency of DNA uptake into the cells can be improved by prior soaking in a solution of calcium ions, e.g. $CaCl_2$. The recombinant plasmid is capable of independent replication in the recipient cells and will confer its genetic characteristics to the host. Thus recombinant pBR322 which has an extraneous strand of DNA incorporated at the PstI site will confer the characteristic of tetracycline resistance but not ampicillin resistance on the host *E. coli* cells which have been transformed.

The cells can be individually selected for these characteristics and grown to high concentrations. Positive identification of clones containing plasmids with the desired DNA inserts can then be made by hybridisation to radioactive samples of the mRNA from which the cDNA inserts were originally synthesized — these radioactive mRNA samples are known as gene probes. However, the presence of the appropriate DNA segment in the plasmid will not result in expression of a protein without the necessary control sequences.

Control sequences which include a promoter, terminator and ribosome binding site must be taken from a well characterized procaryotic gene — the lac or trp

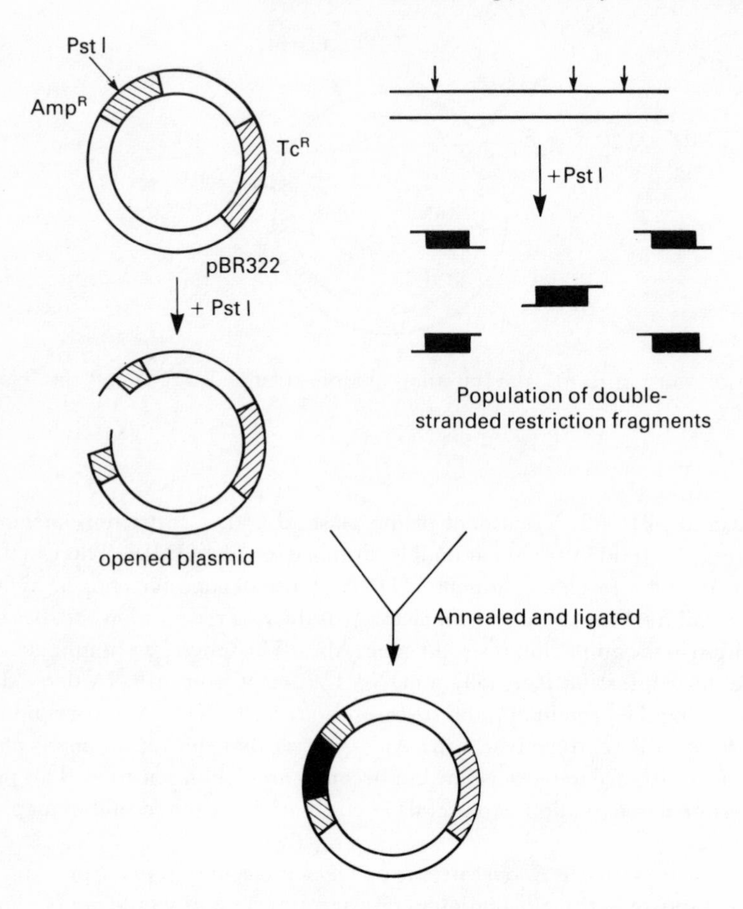

Fig. 3.6 Insertion of isolated mammalian gene into pBR322.

operon control sequences have been extensively used for this purpose. Such control sequences are inserted into the plasmid at appropriate positions relative to the mammalian gene to allow its transcription and translation into a protein. Plasmids constructed with control sequences which may allow an inserted gene to be expressed are particularly useful, and are known as expression vectors. Once a cell has been selected for its expression of a mammalian gene, its growth can be scaled up to the level required. Clearly, the selection process is critical, so that cells with a high specific productivity can be identified and isolated.

Genomic and cDNA Clone Libraries

In order to be able to synthesise a selected protein by recombinant DNA technology, it is necessary to isolate the gene containing the appropriate nucleotide sequence. Such an isolate can be obtained from a population of DNA fragments derived from the total DNA of a cell or tissue of choice. If a restriction endonuclease with a hexanucleotide recognition sequence is used to cleave the extracted nuclear DNA, a population of fragments with an average size of 4 kilobases will be produced. Such fragments would be too small to accomodate an entire gene and so it is usual to carry out a partial restriction digest so that DNA fragments of 10–30 kilobases are obtained. These large fragments may be incorporated into a vector such as a bacteriophage or a cosmid which can be used to transform *E. coli*.

This transformed *E. coli* represents a population of recombinant clones each one of which contains a particular DNA fragment from the original extracted DNA. This *E. coli* population is called a genomic clone library and the individual clones of the library can be screened in order to identify the DNA fragment required.

Screening the genomic library involves the use of a probe which may be a small synthesized sequence of nucleotides representing part of the gene or it may be an mRNA extract from cells that synthesize the protein of interest. The probe is made radioactive by attachment of ^{32}P phosphate on one end by use of an enzyme such as polynucleotide kinase.

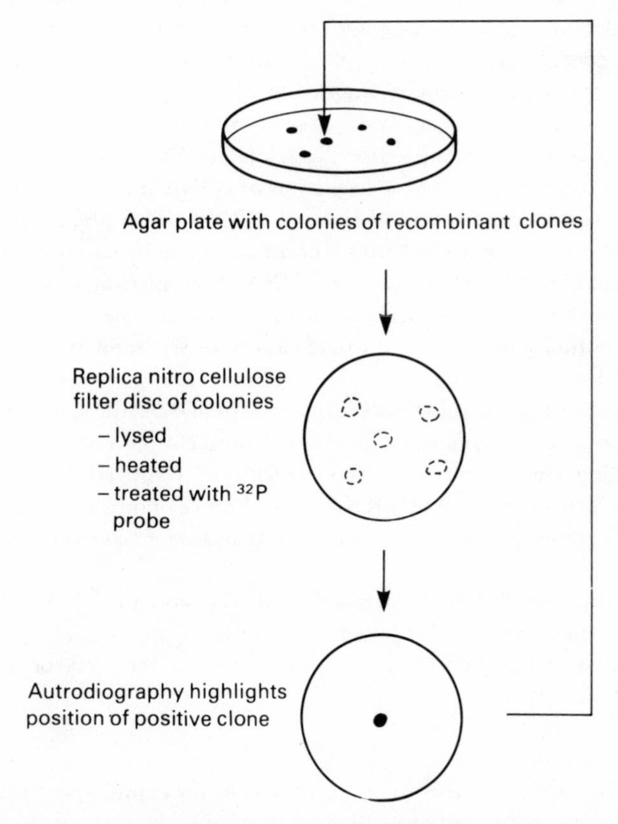

Fig. 3.7 Grunstein-Hogness method of *in situ* hybridization.

The radioactive probe can be used to highlight the recombinant clones containing the required DNA fragment by *in situ* hybridization — a method developed by Grunstein and Hogness (Fig. 3.7). The population of recombinant *E. coli* is inoculated onto a series of agar plates so that individual colonies can form. From each set of colonies a replica plate of the recombinant clones is obtained on a nitrocellulose filter disc. The clones are lysed and fixed on this disc by alkali and heat treatment. The filter can then be washed in a solution containing the radioactive probe which will hybridize to any complementary DNA contained in the clones. Autoradiography of the dried discs will locate the position of the clones with positive hybridization to the probe and this in turn can be used to locate the clones on the original plate.

The number of recombinant clones which must be screened in order to isolate the desired DNA fragment is dependent upon the size of the original DNA and the average size of the fragments produced by restriction cleavage. It can be calculated that to achieve a 95% probability of isolating a particular 20 kilobase fragment from a human genomic DNA library, at least 4.2×10^5 clones should be screened.

It may be that the fragment isolated does not represent the whole of the required gene. In that case, the isolated fragment may be used to re-probe the library of clones in order to obtain overlapping fragments which may extend through the length of the gene. Such a process is called 'chromosome walking' and can be used to obtain a series of overlapping fragments which may collectively contain the whole sequence of a large gene.

An alternative to a genomic clone library is a cDNA clone library. This is a population of recombinant clones derived from cDNA fragments obtained as the product of reverse transcription of isolated mRNA. The advantage of a cDNA library is that it can be enriched with a particular gene by careful selection of the original source of mRNA. For example, cDNA obtained from mRNA of reticulocytes is likely to have a high concentration of the globin gene. This means that the number of recombinant clones that would have to be screened in a cDNA library is considerably less than for a genomic DNA library.

A further advantage of a cDNA library is that the DNA fragments correspond to the coding sequences of genes and do not contain the intervening introns. Introns are non-coding nucleotide sequences which are removed ('spliced') during eucaryotic gene transcription to mRNA. The absence of introns in the cDNA is an advantage for gene expression in bacteria which do not have the mechanism for splicing.

Both genomic and cDNA libraries are a valuable pool for DNA isolation as the first stage of gene cloning and expression of a selected protein. The populations of bacteria represented in the libraries are stable and can therefore be used for gene selection indefinitely.

Eucaryotic Cell Vectors

Viruses can be used as vectors in procaryotic cells for cloning larger fragments of DNA, but this has not found particular application for the expression of mammalian genes. However, the genetic engineering of animal cells is prossible by the use of viruses of which the simian virus — SV40 — has been the most widely used. The double-stranded DNA content of this virus is well characterized as a 5,243 base pair closed circle. It is suitable for genetic manipulation and an extraneous gene may be inserted into the SV40 genome. Addition of eucaryotic control sequences will allow gene expression in the recipient cells. The recombinant virus can be allowed to enter the recipient animal cell by normal viral infection. Alternatively, the naked DNA of the recombinant SV40 can be absorbed into the cells — a process termed transfection, the efficiency of which can be enhanced by calcium ions. The use of naked DNA is preferable because it avoids the need for packaging into the viral proteins which may be inefficient.

The use of naked DNA derived from the SV40 DNA virus is suitable for gene expression in mammalian cells because it has an origin of replication and control sequences which are specific for eucaryotic cells. Depending on the recipient mammalian cell line, the SV40 DNA will replicate as an extrachromosomal element or it may integrate into the cellular DNA.

The same problems of cell selection and gene expression exist with mammalian cell transfection as previously discussed for plasmid insertion in bacteria. Genetic markers are required, and the most useful in mammalian cells are those that do not require the use of a mutant recipient cell line to indicate recombinant DNA acceptance. One such genetic marker which has been used extensively confers resistance to the drug methotrexate, which is a folate analogue and kills cells by binding to the catalytic site of the enzyme, dihydrofolate reductase (DHFR), which is essential for purine and pyrimidine nucleotide synthesis (Fig. 3.8). Resistance to methotrexate may be conferred by the increased expression of DHFR or expression of a mutant enzyme which has a lowered affinity for the drug.

As well as selecting cells that have incorporated recombinant DNA, this DHFR gene can be used to amplify the cell content of the recombinant DNA. Thus, a stepwise increase of the methotrexate concentration of the culture medium will induce the amplification of the DHFR gene. This will involve amplification of the whole recombinant DNA including any linked genes that may have been selected for

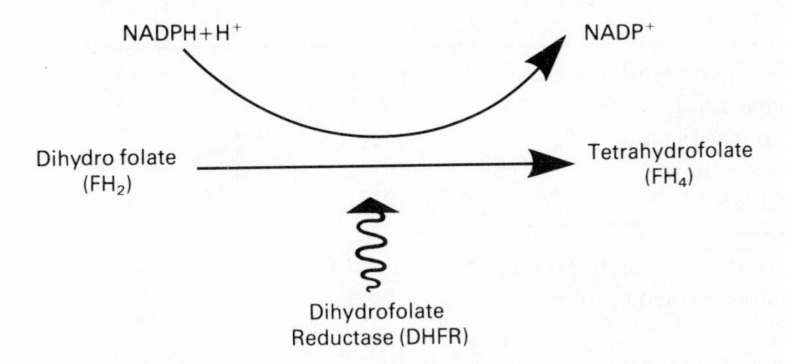

Fig. 3.8 Enzyme system amplified by methotrexate.

protein expression. Such a process of gene amplification may select cells having a high copy number of the recombinant DNA, i.e. several recombinant elements per cell. Such cells may be capable of high specific protein productivity.

Advantages and Disadvantages of Using Genetically Engineered Bacteria for the Production of Biologicals

Considerable progress in the application of recombinant DNA technology was seen in the late 1970s. Several commercially important biologicals were expressed through recombinant plasmid insertion in *E. coli* and many perceived this as heralding the end of the need for mammalian cell cultures. Some of the arguments cited for the use of such genetically engineered bacterial cultures for large-scale production are listed in Table 3.1. The well-documented and simple growth procedures necessary for bacterial cultures are particularly attractive. The media required are usually based on single carbon energy sources and are very much simpler and cheaper than those for mammalian cells, which have supplements of complex growth factors and serum. The growth rate of bacteria is fast — a typical

Table 3.1 Advantages and disadvantages of the use of genetically engineered bacteria

Advantages
High growth rate   doubling time can be ~20 min
Reliable, simple and cheap growth medium
High specific productivity of selected peptides/proteins
Disadvantages

Deficient post-translational modification, e.g.
  proteolytic cleavage
  subunit association
  glycosylation
  acylation

Intracellular location of product   requires extraction from cell lysate
Endotoxins produced   may result in product contamination

doubling time of 15–30 min as compared with 15–30 h for mammalian cells — and such differential growth rates allow higher cell productivities in bacterial cultures (up to × 100).

However, a more sober view of some of the comparisons between the use of genetically engineered bacteria and mammalian cells can reveal many disadvantages for the former. The post-translational modifications associated with mammalian cell proteins do not function in bacteria. These modifications can involve proteolytic cleavage, subunit association or a variety of addition reactions such as glycosylation, methylation, phosphorylation or acylation. Many of these changes are important for the biological function of proteins. For example, glycosylation (which is the most commonly observed modification) can protect proteins against proteolytic breakdown, maintain structural stability and alter antigenicity. This may confer therapeutic advantage to the product. Glycosylated products may be more stable in the blood stream as compared to the non-glycosylated product even though there may be no difference in their activities *in vitro*.

The presence of the Golgi apparatus in mammalian cells allows secretion of proteins which can then increase in concentration in the medium. This enables product extraction to be made from culture supernatants. This is a much more satisfactory prospect than extraction from the mixed protein complement of the product in lysed bacteria. Furthermore, the bacterial lysate may contain endotoxins which must be completely removed before use of the desired protein.

Yeast cells have often been considered as an alternative to bacteria for genetic engineering. They have the advantage of being fast growing cells but with many of the advantages of eucaryotic cells. The only doubt cast over the use of yeast is that although post-translational modifications are made to secreted proteins, some of these modifications may differ from those occurring in animal cells.

The development of genetically engineered mammalian cells is relatively new. Such cells have the advantages of secretion of authentic fully modified products but at much higher specific rates of production than the corresponding non-manipulated cells. Few full scale production processes based on such cells are currently operational but no doubt future developments in this area will afford greater assessment of this approach.

Each of these points of comparison made above and summarized in Table 3.1 varies in importance depending upon the specific product required. In many cases rival processes based on bacterial or mammalian cell systems are now competing and the choice of a preferred method for large-scale production is not always clear. These alternative processes are described for individual biological products in later chapters and assessments made of their relative merits.

Mammalian Cell Hybridization

Another effective method of producing genetically modified mammalian cells is by cell hybridization. This involves the fusion of unrelated cells in suspension by treatment with selected viruses or polyethylene glycol. These procedures were introduced in 1965 by Harris and Watkins who fused human and mouse cells. Initially, the combined cells, called heterocaryons, have two nuclei. Eventually when the heterocaryons proceed to mitosis, a hybrid cell is formed in which the nuclei fuse. Hybrid cell lines can be produced from such a heterogeneous population of cells by cloning. Such cell lines are extremely unstable genetically and tend to lose chromosomes. In the case of mouse-human cell hybrids, the human chromosomes are lost preferentially and this technique has been extremely useful in mapping human genes. Thus, analysis of such hybrids can ascribe particular biochemical functions to particular chromosomes.

The value of cell hybridization was extended further in 1975 by Kohler and Milstein, who produced hybrid cell lines from mouse lymphocytes that were capable of continuous secretion of selected antibodies (see Chapter 6). Despite the lack of the full understanding of the molecular events associated with cell fusion, the technique holds great value in being able to combine the desired characteristics of two independent cell lines by an apparently random assortment of their genes in a population of hybrid cells.

Risks Associated with the Use of Products from Genetically Altered Cells

Many cell products are used for *in vitro* assays — in particular as diagnostic tests. Clearly the criteria for the use of such products are related to the value, precision and sensitivity of the test. However, some of the products of animal cells are required for human injection and often over long periods of time. In this case potential risks such as those listed in Table 3.2 should be considered.

In the 1950s a decision was made that heteroploid or tumour cells were unacceptable for products destined for human consumption (see Chapter 1). Since then, a number of important biologicals have been produced from such animal cell lines and the original safety criteria questioned. Similarly, arguments for the use of

Table 3.2 Potential risk factors associated with the use of animal cell products

Risk factor	*Effect*
Cell	Tumour
Proteins	Immune response transformation
Nucleic acids	Integration and expression of abnormal genes
Endogeneous viruses	Transformation

(From Petricciani, 1985)

products from genetically engineered bacteria or animal cells are being advanced.

The major problems concerned with the use of these cells relate to the possibilities of contaminating the products with factors that might induce tumours in the recipient. Stringent purification techniques can limit the presence of most of these factors. However, of greatest concern is the assessment and monitoring of the risk of transmitting fragments of DNA that might contain oncogenes with the associated risk of cell transformation in the recipient. This concern is expressed in a background of only limited understanding of the process of tumorigenicity.

However, the risk of tumorigenicity by the residual cellular DNA in the product can be assessed from the minimum detection levels for DNA that can be assayed in the product compared with the minimum level of oncogenes required for cellular transformation. Current methods of DNA assay by hybridization can determine levels as low as 1 pg/ml. This minimum residual DNA concentration would represent a random assortment of DNA fragments from the cellular genome. From such a base level of residual DNA, the whole unfragmented oncogene content can be assessed as unlikely to be above 10^{-6} pg of DNA. This should be compared with the minimum oncogenic level of DNA required for cellular transformation of 100 pg which is an order of 8 logs higher.

Such evaluations relating the safety of a cell product to its standard of purity and minimum levels of contamination are currently being expressed as arguments for the use of heteroploid or tumour cell lines for the large-scale production of biologicals. These arguments are compelling, especially when considered alongside the medical/therapeutic benefits of many of the compounds under discussion. The debate is unlikely to be resolved in terms of a blanket approval for the use of these cell lines for the production of such compounds — rather each biological product from such cell lines will be assessed in terms of individual risk/benefit ratios. Approval and licensing for the large-scale use of such compounds would then be determined for each product independently by national regulatory authorities.

Summary

The use of certain purified microbial enzymes — notably restriction endonucleases, ligases and reverse transcriptase — has lead to the development of recombinant DNA technology which has allowed the manipulation of genes. Such manipulations can be used to insert a mammalian gene into a vector which can be incorporated into a suitable recipient cell. Clones of such recipient cells can be selected by means of genetic markers and cultured to allow the expression of the incorporated gene. Such expression will only occur if the appropriate control sequences are present in the recombinant vector.

Mammalian gene expression using plasmids in *E. coli* has been particularly successful in the production of a range of mammalian cell products. In contrast, gene manipulation in mammalian cells is in its infancy, but promises to be a useful means of generating valuable cell lines in the future.

Despite the apparent success of recombinant DNA technology in generating bacterial cells capable of synthesizing mammalian proteins, there are a number of disadvantages to this approach which relate to the lack of eucaryotic enzymes in these cells. This causes problems with certain products which require modification by these enzymes before they are biologically active. This suggests the preferential use of animal cells which may be genetically modified for the production of selected mammalian proteins.

Cell hybridization results in the random assortment of genes from two independent cell lines. This has proved valuable in mapping human genes as well as generating cell lines capable of secreting monoclonal antibodies with a predetermined specificity.

The question of the safety of products from genetically modified cells is particularly pertinent for those destined for human consumption. The danger of contamination with tumorigenic DNA must be carefully assessed. At one time only 'normal' diploid human cells could be considered for such production. However, increased product purification and increased sensitivity of DNA monitoring may warrant an acceptance of such products if they are of medical benefit.

General Reading

Brown, T.A. (1986). *Gene Cloning: an introduction.* Locahin van Nostrand Reinhold.

Emery, A.E.H. (1984). *An introduction to Recombinant DNA*. Chichester, J. Wiley & Son.

Murray, K. (1980). 'Genetic engineering. Possibilities and prospects for its application in industrial microbiology', *Phil. Trans. Royal Soc. London B* **290**, pp. 369–386.

Old, R.W. and Primrose, S.B. (1985). *Principles of Gene Manipulation*. Oxford, Blackwell Scientific.

Watson, J.D. *et al.* (1983). *Recombinant DNA: a short course*. Reading, W.H. Freeman & Co.

Specific Reading

Beale, A.J. (1979). 'Choice of cell substrate for biological products', *Arch. Exp. Med. Biol.* **118**, pp. 83–97.

Petricciani, J.C. (1985). 'Regulatory considerations for products derived from the new biotechnology', in *Large-scale Mammalian Cell Culture*. Eds. Feder, J.B. and Tolbert, W.R. pp. 79–86. London, Academic Press.

Spier, R.E. (1982). 'Animal cells or genetically engineered bacteria for the manufacture of particular bioproducts', *Develop. Biol. Stand.* **50**, pp. 311–321.

Weymouth, L.A. and Barsoum, J. (1986). 'Genetic engineering in mammalian cells', in *Mammalian Cell Technology*. Ed. Thilly, W.G., pp. 9–62. London, Butterworth.

SECTION II

Animal Cell Products

Chapter 4

Vaccines

Introduction

The large scale use of vaccines as prophylactics has been one of the greatest achievements in disease control over the last 50 years. Vaccines may now be produced for protection against a variety of viral, bacterial or parasitic diseases. It is the development and production of viral vaccines which will be considered in this chapter because of their close association with animal cell cultures. The importance of these viral vaccines in the improvement of world public health can be highlighted by the complete eradication of smallpox and the virtual eradication of poliomyelitis. Both of these were at one time widespread diseases, particularly amongst the young.

The concept of vaccines dates back to the late eighteenth century with Edward Jenner's observation that farm workers who had contact with the bovine disease of cowpox developed immunity against smallpox. The cowpox virus or vaccinia (from the Latin 'vacca' meaning cow) lent its name to vaccination which was the deliberate injection of the virus so as to protect against the effects of smallpox. This practice was developed in the nineteenth century when the virus was obtained from infected cows. The live vaccinia virus was used as a smallpox vaccine up to the time of eradication of the disease in 1978.

In 1885, Pasteur reported the development of another human viral vaccine — for rabies. The human disease is normally contracted following a bite from an infected dog. The virus spreads slowly along the nerves and several weeks may elapse before the full disease symptoms develop. Because of this relatively long incubation period, the rabies vaccine is unusual in so far as it can be administered after viral infection, i.e. after the victim has been bitten. Pasteur produced his vaccine by propagating the rabies virus in rabbits. The spinal cord of an infected rabbit was removed and suspended in sterile air in which it was dried (over potassium hydroxide). Successive injections of fragments of such infected spinal cords were used to successfully treat several victims of bites from rabid dogs.

Jenner's smallpox vaccine and Pasteur's rabies vaccine were the only two human viral vaccines produced before the 1930s. The difficulty for further development of vaccines arose in finding a suitable host other than whole animals. In the 1930s fertile hen's eggs were used for the propagation of several viruses. The chick embryos contained in the eggs have several advantages over whole animals for virus production. They are cheaper, easier to handle and less liable to cross-contamination. Yellow fever vaccine was produced for the first time in the 1930s by this means.

It was the discovery by Enders in 1949 that poliomyelitis virus could be grown on primary cells in culture that led to the modern era of vaccine production. In fact, the major innovations in animal cell technology which were initiated in the 1950s resulted from the drive towards a mass vaccination campaign particularly against polio.

Principle of Viral Vaccines

Viruses are microscopic particles ranging in size from 10 to 500 nm. They consist of a nucleic acid core (which may be RNA or DNA) surrounded by a protein coat. They cannot be propagated on their own but depend upon animal cells whose protein and nucleic acid synthetic capabilities can be controlled and utilized for further virus production. The genetic complement of a virus is simple. It consists of the coding sequences for the coat protein and the control sequences utilized in the host cell.

Many viruses are capable of replication in animal cells, after which they may be released by lysis or degeneration of the host cell. The typical events associated with a viral infection of this type are shown in Fig. 4.1. This lytic cycle consists of distinct steps of adsorption, penetration, replication and release. During the cycle the virus controls the cellular metabolism leading to the formation of new virus particles (virions) and lysis of the cells. The infection of an animal by a virus elicits antibody production which can reduce the spread of the infection and protect against future infections. The protective antibodies are produced against antigenic determinants contained in the viral coat protein.

The basis of a vaccine is to allow exposure to these antigenic determinants without producing the undesirable disease symptoms. The three major vaccine types which allow this are:

(1) Viable disease-associated viruses which are inactivated
(2) Live viruses which are modified or attenuated so that they retain their antigenic characteristics but lose their pathogenicity i.e. they cannot induce the disease
(3) Peptides containing the antigenic determinants — these can be isolated from the virus, produced by genetic engineering or chemically synthesized

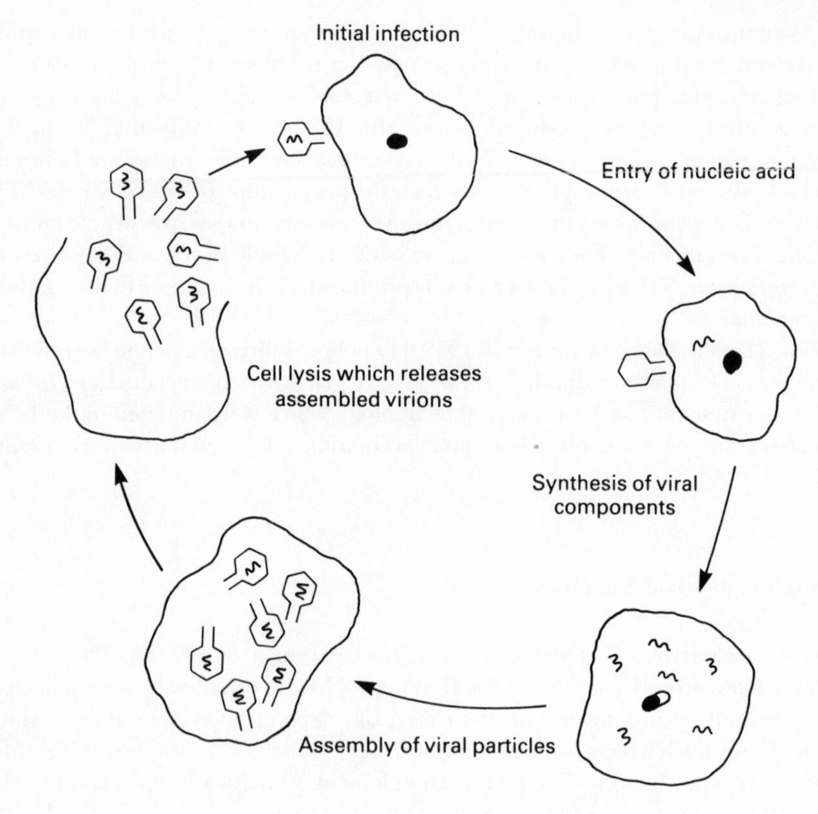

Fig. 4.1 The lytic cycle of viral infection.

Production of Inactivated Viral Vaccines

The propagation of viruses in cultured cells was established as a basic technique in the 1950s and has been amenable to scale-up for large-scale vaccine production. The use of multi-roller bottle or microcarrier cultures allows the production of large quantities of anchorage-dependent cells which can be infected by the appropriate virus (see Chapter 2). Viral propagation continues until a high titre is observed in the culture. The harvested virus is then treated by an inactivating agent such as formaldehyde which renders the virus non-pathogenic and suitable for use as a vaccine. The development of such vaccine production methods from the laboratory to the industrial scale can be illustrated by two specific examples.

Poliomyelitis Vaccine

In 1954 Salk developed the inactivated polio virus vaccine. This vaccine is still in extensive use and the basic stages of its production are as follows:

(1) A primary cell culture is established by trypsinization of an isolated monkey kidney
(2) The cells are grown to a confluent monolayer. The original method for this was the roller bottle system. However in many large-scale vaccine plants this has now been replaced in favour of the microcarrier culture system which offers advantages in scale-up
(3) The virus is inoculated into the confluent cell culture in which it propagates until a maximum virus titre is obtained. Typically this would take about 3 days
(4) The culture medium is then concentrated and the virus purified. Concentrations of up to × 250 may be achieved by ultrafiltration. Purification can involve gel and ion exchange chromatography to ensure the elimination of any traces of serum protein used in the culture
(5) The virus is inactivated by treatment with formaldehyde
(6) The vaccine is formulated so as to include samples of the three major strains of poliovirus. Inactivated preparations of all three strains are combined in a pre-determined ratio to produce the final trivalent vaccine. An adjuvant such as aluminium hydroxide may also be added. The purpose of this is to stimulate the immune response

Significant modifications were made to the early production procedures as a result of later events:

(1) The primary monkey kidney cells first used as a substrate for polio virus propagation caused some concern with the discovery of an associated simian tumorigenic virus — SV40 — which was present in the cells and found to contaminate the vaccine. This led to concern about the origin of cells to be used as substrates for human injectable products. It was considered that the only cell types suitable were normal human diploid cells or primary cells taken from animals with no demonstrable viral infection. This led to the widespread use of low passage number human diploid fibroblasts which were cultured from tissue biopsies. In particular, two well defined cell lines — WI-38 and MRC-5 — derived from human embryonic tissue were found suitable for vaccine production. These cells have limited life-spans but cultures established from embryonic cells have sufficient growth potential for the needs of vaccine production (see Chapter 1).

    Recently, the African green monkey cell line, Vero, has been used as a cell substrate for polio virus propagation. This is a continuous cell line with good growth characteristics. Despite its heteroploid genetic character it has not been shown to be tumorigenic and vaccine production from this source has now been accepted
(2) The occurrence of several instances of vaccine-associated polio, following field trials in California, necessitated further modification of polio vaccine production. Residual viable virus activity was found in some batches of the administered vaccine and an inefficient de-activation process was implicated

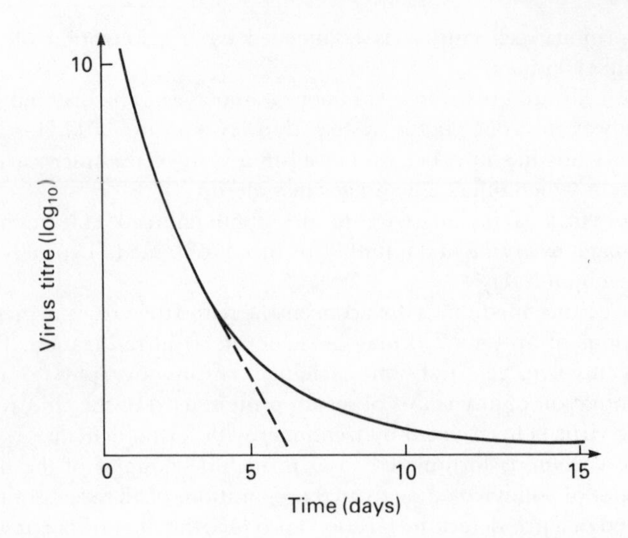

Fig. 4.2 A typical survival curve of a virus during formaldehyde treatment. The graph shows the shape of the plot of the logarithm of the plaque forming units of virus against the length of formaldehyde treatment. The dotted line represents a mistaken linear extrapolation of the survival curve to zero.

The length of time of formaldehyde treatment required for complete inactivation of the virus was based on the extrapolation of virus titre survival curves to zero activity (Fig. 4.2). However, it would seem that a tailing effect occurs at low virus titres which indicates resistance of low viral concentrations to inactivation. Many explanations for this have been proposed. These include the aggregation of some virus particles into resistant clumps. Formaldehyde reacts with both the protein and nucleic acid content of the virus. It is suggested that the virus particles may become increasingly impermeable to formaldehyde as the protein coat is 'locked up' and thus preventing reaction with nucleic acid. Current methods of viral inactivation still involve formaldehyde treatment but for considerably longer incubation times than were originally used.

Foot-and-Mouth Disease (FMD) *Vaccine*

Because of the widespread economic effects of this disease on agricultural production, this vaccine is now produced in considerable quantity. In cattle-producing countries where the disease is endemic, two or three injections of vaccine per cow per year may be administered. This leads to a world requirement of 1.5×10^9 doses per year which exceeds the production level of any other cell product.

Originally blood extracts from infected animals were used as a source of vaccine. This was later superceded by a process of virus propagation on bovine tongues obtained from slaughterhouses. Subsequently, the FMD virus was grown on Baby Hamster Kidney (BHK) cells established as a cell line suitable for continuous growth in fermenters. These cells are anchorage-dependent but can be adapted for growth in suspension culture. The method of BHK culture (i.e. anchorage or non-anchorage) depends upon the strain of FMD virus to be cultivated. Currently, batch cultures up to 4000 l are being used in such FMD vaccine production.

In the production of veterinary vaccines, safety regulation are more relaxed. Transformed animal cells have been accepted as viral substrates which makes culture much easier particularly when non-anchorage dependent suspension cells are used. Aziridines are used as agents for inactivation. They do not show the 'tailing effect' of inactivation exhibited with formaldehyde although their potential carcinogenicity rules them out in processes for human vaccine production.

Two other widely used inactivated viral vaccines deserve comment. The rabies vaccine is used extensively as a veterinary vaccine in parts of the world where the disease is endemic. As a human vaccine its main use is in post-exposure treatment although it can be used selectively to vaccinate recipients with a high risk of exposure. Inactivated influenza vaccines are also generally available. However the antigenic determinants of each strain of influenza virus shows considerable genetic variability. This reduces the effectiveness of the vaccine which must be matched to the strain causing the prevailing epidemic.

Attenuated Viral Vaccines

An attenuated viral vaccine consists of a disease-causing virus which has been genetically modified so that its antigenicity is preserved but its pathogenicity is reduced. The attenuation can be induced by mutagenic agents but in most cases it is allowed to occur spontaneously by continuous passage of the virus in cell culture. The advantage of a live viral vaccine is that it is capable of multiplication in the injected host, who then receives a much greater immunogenic stimulation than from an equivalent injection of dead virus. The problems of residual infectivity encountered with the normal virus do not occur, although the potential for genetic reversion of the attenuated virus must be considered.

Vaccinia is a naturally occurring attenuated virus used as a vaccine against smallpox. Antibodies induced by the vaccinia virus are also effective against the smallpox virus. The use of this vaccine has led to the complete eradication of the disease from the world population.

In 1962 Sabin developed a live attenuated polio vaccine which subsequently found widespread use. The vaccine can be administered orally. The advantages considered for the introduction of the live Sabin polio vaccine included the following:

(1) It produces a local cellular resistance of the walls of the intestinal tract as well as producing antibodies in the circulating blood — a feature of oral administration
(2) It may be passed on to contacts of those vaccinated, thus leading to the possibility of community protection and the eventual eradication of the viral disease
(3) During an outbreak of the disease, the attenuated virus can compete with the wild type in the alimentary tract and so exert a protective effect prior to the generation of antibodies

(4) The dose required for immunization is a factor of $\times 10^5$ less than that required for the inactivated vaccine. This particularly affects the extraction process from cell cultures. The cell culture medium requires dilution for preparation of the attenuated viral vaccine whereas the culture medium requires concentration for preparation of the inactivated vaccine. This means that the cost of production of the attenuated vaccine is much lower.

Both the Salk and Sabin polio vaccines were introduced after serious epidemics of the disease in the early 1950s. Since their introduction the incidence of the disease has decreased in the industrialized nations from 100–200 polio cases per million of the population to one case per 4.5×10^6 doses of vaccine. Although clearly successful in reducing the incidence of polio, the arguments originally used for the relative merits of the two vaccine types have changed. There are now a number of arguments for the re-introduction of the Salk vaccine in countries such as the UK where it was replaced by the Sabin vaccine. Some of these points are particularly relevant to Third World countries where the disease incidence is much higher. These arguments include:

(1) Now the overall incidence of poliomyelitis is extremely low, the risk of contracting the disease from the vaccine becomes significant compared to infection from wild strains. Of the small number of incidents of polio, some of these can be ascribed to reversion of the attenuated virus strain to a virulent form. Recipients with a low immune response, often found in Third World countries, may be particularly susceptible to such vaccine-induced disease. The inactivated vaccine has no such associated risk
(2) As cell culture techniques have improved, the cost difference between the production of the two vaccine types is not as significant as when they were first introduced
(3) Contrary to the original arguments, countries such as Finland and Sweden have been able to completely eradicate poliomyelitis by the sole use of the inactivated Salk virus

The history of the development and use of the polio vaccines provides a very good example of how the balance of arguments for and against a particular product may alter with a changing technology and changing circumstances. It is predicted that such arguments will have to be further balanced in the future with an increasing number of vaccine types becoming available for each disease.

Despite the arguments used above, live attenuated viruses have been the preferred vaccine type for many diseases for which an identifiable virus can be grown in cell culture. Some inactivated viruses may have to be administered in multiple doses and this can be expensive. In some cases such as Foot-and-Mouth disease, suitable attenuation of the pathogenic virus has not proved successful and so the inactivated virus has been used. In other cases, the inactivation process may cause unacceptable loss of antigenicity. For several human diseases — notably mumps, measles and rubella — attenuated viral vaccines grown on embryonic chick cells

Table 4.1 Some routinely used whole virus vaccines

Human	*Veterinary*
Polio (Salk)*	Foot-and-Mouth disease*
Polio (Sabin)	Marek's disease
Measles	Newcastle disease
Mumps	Rinderpest
Rubella	Rabies
Yellow fever	Canine distemper
Rabies*	Swine fever
Influenza	Blue tongue
	Fowl Pox

Those indicated * are normally produced as inactivated viruses. All the others are normally produced as live attenuated viruses.

in culture have been highly successful in significantly reducing the disease incidence in the industrialized world.

Table 4.1 shows examples of some of the most widely used human and veterinary vaccines which are produced from whole viruses.

Subunit Vaccines

An isolated or synthesized portion of the viral coat protein can be used as an alternative to a whole virus (attenuated or inactivated). However, development of alternatives to the traditional viral vaccine production methods established in the 1950s has been slow. There are various reasons for this based on economic considerations:

(1) The production costs of whole virus vaccines are relatively low. Subunit vaccines require extraction, purification and inactivation. Each injected dose of the isolated antigen needs to be up to $\times 10^5$ greater than the equivalent live virus vaccine
(2) The quality controls and testing required for the introduction of a new vaccine is extensive. This would require further development costs
(3) The development costs are too high compared to the scale of use of many vaccines. In many cases, a vaccine may be directed to Third World countries which do not have the financial resources required for these developments

The greatest impetus for the development of alternative vaccines has been in cases where the virus is not easily propagated in cell culture. One important example of this is hepatitis B. Although the virus associated with this disease has been well characterized, it has not been found possible to propagate it in cell cultures. The immediate prospect for vaccine development has been to use extracts from the blood of chronic carriers. Pooled plasma from such carriers contains the complete virus as well as spherical particles consisting of viral coat protein. These spherical particles which are 22 nm in diameter are entirely composed of viral protein and are antigenic. These protein subunits of the virus can be extracted and purified from the pooled plasma and then used as a vaccine. Providing the final extract is free from nucleic acid contamination this has been found satisfactory in protecting recipients from the disease.

However there are a number of difficulties associated with this process:

(1) Large quantities of pooled plasma are required from persistent carriers. Such plasma is difficult to characterize and suffers from variation between batches
(2) The supply of the plasma may be difficult in the long term as the incidence of the disease decreases
(3) Other unidentified viruses may be associated with the pooled plasma and may be carried over to the extracted product, e.g. HIV — the virus associated with AIDS
(4) The extraction procedure requires the use of large virus containment facilities because of the infectious nature of the plasma

These difficulties in preparation of the hepatitis B sub-unit vaccine makes the process expensive and the long term supply limited.

Synthetic Polypeptide Vaccines

Considerable advances have been made in identifying the precise antigenic region of the protein surface of certain viruses. This identification can be accomplished by a trial and error approach of testing fragments of viral coat protein for immunogenicity. The success of this approach can be enhanced by prior location of regions of the coat protein with a high concentration of hydrophilic groups. These groups are likely to form the outer surface of the virus and play a major part in antigenicity. Such predictions of hydrophilicity can be made from amino acid sequence data and computer predictions of secondary and tertiary structure.

Small peptides of eight to ten amino acids have been isolated by this technique from Foot-and-Mouth disease and polio viruses. The peptides correspond to regions of the viral protein which are highly variable particularly between viral sub-types. These viruses were chosen because of their relatively simple structures. Although these peptides are immunogenic and could serve as effective vaccines, they are unlikely to be commercially viable for use against these particular diseases because of the ready availability of cheap and effective vaccines produced by the traditional methods.

The development of synthetic peptide vaccines against hepatitis B is of greater importance because of the difficulties associated with vaccine production for this disease as outlined earlier. It has been possible to clone and sequence the whole genome of the hepatitis B virus in bacteria and yeast.

From such clones the whole viral coat protein can be synthesized. This approach has been found to be particularly successful in yeast, where the proteins associated with the original virus can be glycosylated and assembled to form a particle of similar characteristics to the 22 nm spherical subunit extracted from human carriers. These hepatitis B surface antigen (HBsAg) particles can be produced in large quantity from yeast. They have shown good immunogenicity in animals and have potential for use as a human vaccine.

Some success in developing a small peptide vaccine for hepatitis B has been found with the isolation of specific antigenic sequences of 15 to 25 amino acid residues. Cyclization of the peptide and attachment to a carrier protein has been shown to increase the immunogenic response. Although this small peptide vaccine may be less immunogenic than the subunit particles prepared from human plasma, there are a number of advantages:

(1) The peptide can be produced simply by molecular cloning or by solid-phase chemical synthesis. Either of these methods is suitable for large-scale production
(2) It can be prepared in a pure state thus avoiding any side-reactions caused by the presence of impurities
(3) The production process is relatively safe and does not require the same containment facilities necessary in the isolation of subunit particles

Recombinant Live Vaccines

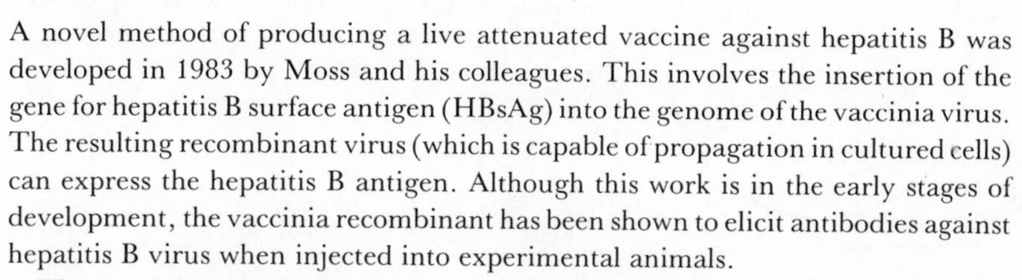

A novel method of producing a live attenuated vaccine against hepatitis B was developed in 1983 by Moss and his colleagues. This involves the insertion of the gene for hepatitis B surface antigen (HBsAg) into the genome of the vaccinia virus. The resulting recombinant virus (which is capable of propagation in cultured cells) can express the hepatitis B antigen. Although this work is in the early stages of development, the vaccinia recombinant has been shown to elicit antibodies against hepatitis B virus when injected into experimental animals.

The vaccinia genome is too large (~ 180 kilobases) for straightforward manipulations by recombinant DNA techniques. Consequently the construction of the recombinant virus involved several steps (Fig. 4.3):

(1) The fragment of DNA known to express hepatitis B surface antigen was isolated with the restriction endonuclease, BamH1
(2) This fragment was fused into a gene for the enzyme, thymidine kinase (TK) which was previously isolated from the vaccinia genome. The fusion forms a recombinant DNA fragment in which the HBsAg gene replaces the TK gene but the promoter region of the latter remains. This promoter region is essential for the expression of any gene
(3) The newly formed recombinant plasmid and the vaccinia virus were allowed to co-infect a monolayer of cultured cells

(1)

1.35 kilobase DNA fragment form the hepatitis B virus.

(2)

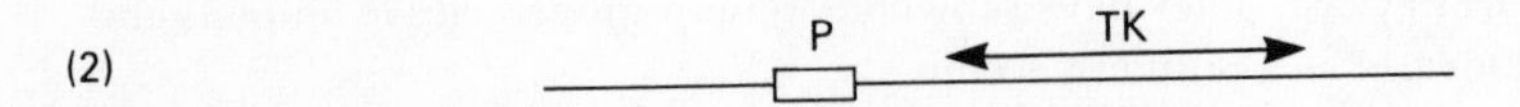

An opened plasmid containing a promoter (P) and TK gene originally isolated from vaccinia.

(3)

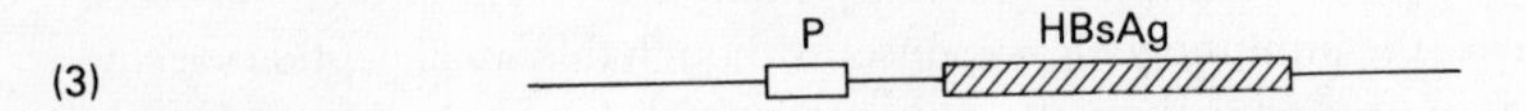

Insertion of the HBsAg gene into the TK gene location.

(4)

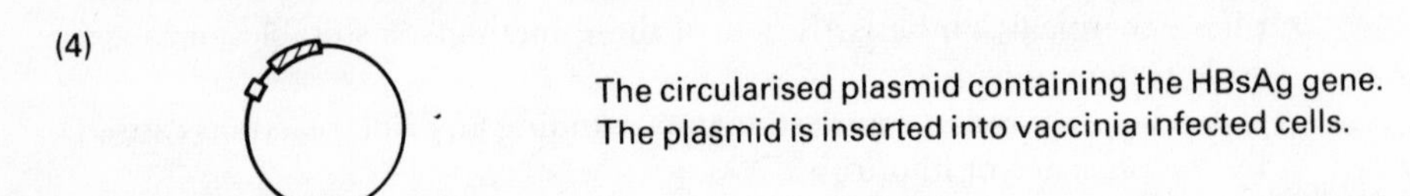

The circularised plasmid containing the HBsAg gene. The plasmid is inserted into vaccinia infected cells.

(5)

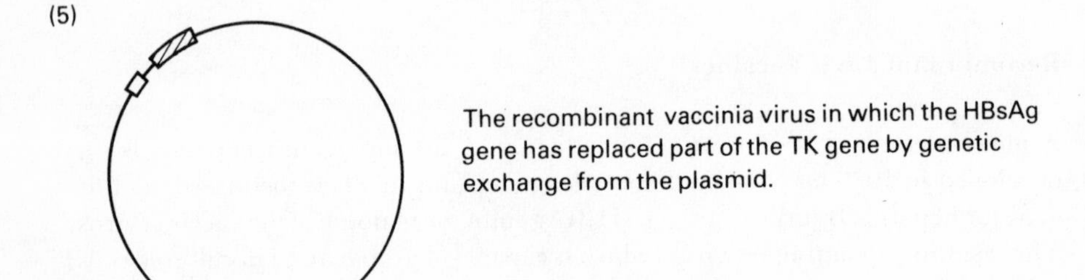

The recombinant vaccinia virus in which the HBsAg gene has replaced part of the TK gene by genetic exchange from the plasmid.

Fig. 4.3 Formation of a recombinant live vaccine.

(4) Propagation of the plasmid and virus in the cells allows genetic recombination by which the viral TK gene may be replaced by the recombinant plasmid. This results in the inactivation of the TK gene

(5) The TK⁻ virus was selected following growth on TK⁻ human cells in a medium containing 5-bromodeoxyuridine (BUdR). This compound is normally toxic to cells. In the presence of the TK enzyme BUdR is converted to its triphosphate which can act as an analogue to dTTP. Its incorporation into DNA prevents correct development of the virus. The absence of the TK enzyme in the cell prevents the metabolism of BUdR and so is no longer toxic. Some of the viruses selected by this means were

shown to hybridise with a ^{32}P-labelled HBsAg gene probe

(6) Further propagation of the isolated recombinant viruses has shown their ability to express the HBsAg protein which can elicit the appropriate antibodies when injected into rabbits

Flanking the HBsAg gene with the TK gene sequences has two purposes in this procedure. Firstly, it promotes recombination at the site of the vaccinia TK gene which is not required for viral infectivity. Secondly it produces TK^- viruses which are easily selected in the BUdR containing media. Expression of the HBsAg gene is controlled by the TK promoter which is retained in the vaccinia genome.

Although this type of recombinant vaccine has not yet been used in humans, it has many promising possibilities. It can increase the potency of a polypeptide antigen by combining the advantages of a live vaccine which is effective at lower dosages.

The isolation and purification of HBsAg required from the blood of infected humans or from the media of genetically engineered cells is tedious and expensive. Such purification procedures are not necessary in the preparation of the recombinant vaccine. The vaccinia virus is chosen for such recombinant work because it is known to be safe for human injection. It has been used widely and successfully in protection against smallpox and any dangers of reversion to a virulent form are minimal. This technique has been used with varying levels of success with antigens of the influenza, herpes and HIV viruses.

Further development suggests the possibility of the construction of a polyvalent vaccine in which several unrelated polypeptide antigens may be expressed in a single recombinant vaccinia virus. This could lead to the administration of several antigens in a single dose of a carefully constructed live multi-vaccine.

The Use of Antibodies as Vaccines

Passive Immunization

Passive immunity against disease may be attained by administration of foreign antibodies that themselves protect the recipient from the disease. The antibodies used can be purified immunoglobulins from blood donors who have been exposed to the disease. Heterologous serum (from a horse or cow) may be used, but this may result in undesirable side effects to the recipient. Monoclonal antibodies from suitably constructed human hybridomas are preferable. This allows the possibility of large scale production of a highly specific antibody (see Chapter 6).

Passive immunization by intravenous injection of antibodies offers immediate protection but is short-lived. The half-life of immunoglobulins in the blood stream is about 3 weeks and protection may last at most for a few months. This contrasts the active immunity offered by virus-prepared vaccines, which can allow protection for several years if not a life-time.

Anti-Idiotype Antibodies

The extensive development of antibody technology (see Chapter 6) has allowed the production of a novel type of vaccine which is not yet commercially available but is likely to provide a major contribution in the future.

An anti-idiotype antibody is one that is produced against another antibody and is specific for the variable region of that antibody. Consider the injection of an antibody (Ab-1) into a rabbit (Fig. 4.4). This will elicit the production of a population of antibodies which bind with various specific sites of Ab-1. Many of these antibodies will react with sites associated with the constant regions of the heavy or light chains. However, some may be specifically produced against the variable region which has a steric arrangement unique to Ab-1. Isolation of such a second antibody (Ab-2) will have a variable region complementary to that of Ab-1. The nature of such complementary binding suggests that the original antigen which complements the binding structure of Ab-1 must have an identical structure to Ab-2 around the binding site. Thus the administration of Ab-2 (the anti-idiotype antibody) to a recipient may be equivalent immunologically to the injection of the original antigen.

The notion that such an anti-idiotype antibody could serve as a vaccine against a viral disease was originally proposed by Nisonoff and his colleagues. Experimental verification of this proposal was attempted in 1986 by Dreesman who injected human antibodies against hepatitis B surface antigen (HBsAg) into rabbits. This anti-idiotype antibody was isolated by a series of immunoaffinity columns which rejected all unwanted antibodies. The resulting anti-idiotypes were shown to compete with HBsAg for the binding site of the human antibodies. Injection of the anti-idiotypes into experimental animals has shown that the same immunological response is produced as compared with injection of HBsAg. Further, protection against the pathological effects of hepatitis B virus by the anti-idiotype vaccine was shown in chimpanzees who suffer the same disease symptoms as humans. The large scale production of such a vaccine could be performed by immortalization of lymphocyte cells producing the anti-idiotype antibodies (see Chapter 6).

Although the technology of anti-idiotype vaccines is in its infancy, the future potential is considerable. This is particularly so for the production of vaccines against diseases for which the pathogenic agent is difficult to characterize or where the antigenic components may be too large for the techniques of recombinant DNA. The advantages of such vaccines include their lack of potential pathogenicity and their specificity which should prevent any unwanted side-reactions. Furthermore, the ability to produce high yields of antibodies from the growth of cultured hybridomas should lead to a relatively low cost product.

Conclusion

The techniques developed in the 1950s for the growth of viruses in cell culture established an extremely valuable technology for the production of vaccines. The large-scale production and administration of these vaccines has led to the eradica-

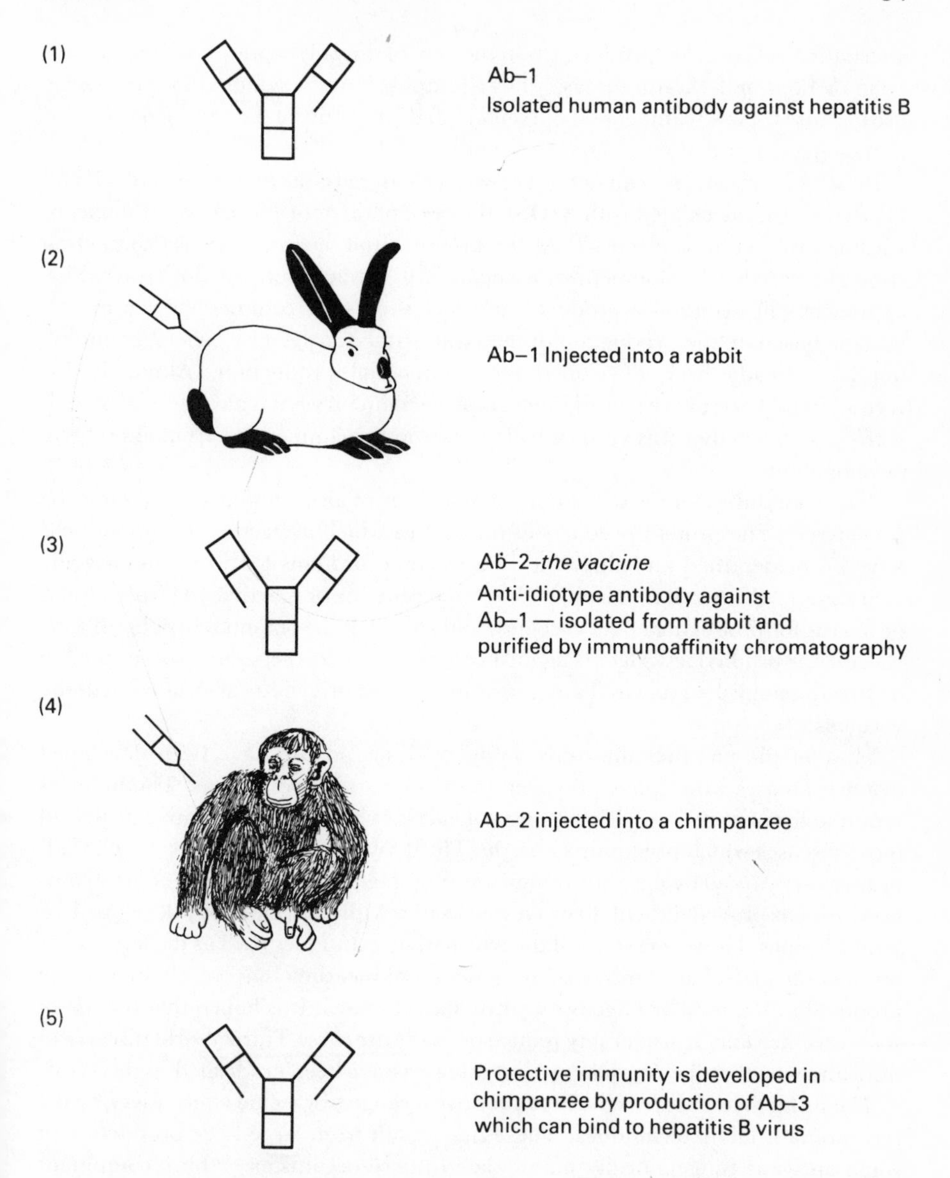

Fig. 4.4 The formation of an anti-idiotype vaccine.

tion or marked reduction of many human diseases. The production methods used are cheap and effective. Even though there may be discussion over the relative merits of inactivated vs live attenuated viral vaccines, the technology for the routine production of each is similar.

The effectiveness of these production methods has been largely responsible for the lack of impetus for the development of alternative vaccine types. Even when alternative vaccines are produced as in the case of the polypeptide antigens against polio or Foot-and-Mouth disease, they are unlikely to be adopted for large-scale use because their advantages do not outweigh the benefits of the experience gained in the effective use of the original vaccines.

In cases where viruses can not be grown easily in cell culture, e.g. hepatitis B and HIV (the virus associated with AIDS), the development of alternative strategies of vaccine production is essential. At the present time there are several competing strategies for the development of a hepatitis B vaccine but it is not clear which approach will eventually produce the most successful commercial vaccine. A vaccine based on the production of the viral surface antigen by genetic engineering has already been developed for commercial production. Although this recombinant vaccine is a considerable improvement on what is already available, it is to be seen whether this vaccine will withstand the competition from alternative developments.

The possibilities for the development of a range of anti-tumour vaccines may be considered. The protein products of oncogenes which are active in tumour cells have been identified and shown to be similar to proteins found in tumorigenic viruses, e.g. Epstein Barr virus (EBV). Such proteins or inactivated viruses could be a basis for producing novel vaccines, and although the potential benefits of such vaccines as prophylactic agents against cancer are clear, the safety aspects need to be firmly established to avoid any possibility of tumorigenesis associated with the vaccines.

Most of the vaccines that have found widespread use have been developed against diseases which have prevailed in the industrialized nations. The finances required for research and development of vaccines has led to the relative neglect of those diseases which predominate in the Third World. Many of these 'neglected' diseases are caused by parasitic organisms (e.g. malaria) and because of their complexity it has proved difficult to produce vaccines with the technology developed for viral vaccines. However, some of the alternative techniques such as the large-scale production of isolated antigens by genetic engineering may result in vaccine production that may be effective against such diseases. It is hoped that the use of such vaccines may considerably reduce or eradicate these Third World diseases in the same way that the prominent viral diseases have been eradicated in the West.

The immediate future is likely to produce a range of vaccine alternatives by the recombinant DNA techniques. These may result from large-scale production of small antigens by genetically manipulated micro-organisms or by recombinant vaccinia viruses. However, hybridoma technology has also developed at an extremely rapid pace in the last few years and may be able to contribute to vaccine production with the anti-idiotype antibodies. The range of these alternative technologies makes it difficult to predict with certainty what the prevailing productions methods will be in the long-term.

Summary

Despite the developments of the smallpox vaccine in the 18th century and the rabies vaccine in the 19th century, the large-scale production of many human viral vaccines did not occur until the advent of routine animal cell culture technology in the 1950s. The ability to propagate viruses in these cultures led to the commercial production of the inactivated and the live attenuated viral vaccines. The widespread use of these led to the rapid decline or virtual eradication of many human and veterinary diseases.

Alternative strategies of vaccine production have developed from the ability to isolate the regions of the viral coat protein responsible for antibody production. These antigenic determinants have proved in many cases to be small peptides of a sequence of around 20 amino acids. The analysis of such peptides can lead to their rapid production by chemical synthesis or by recombinant DNA techniques using genetically engineered micro-organisms. Recombinant DNA technology can also be applied to the production of a recombinant vaccinia virus bearing a gene for an isolated antigenic determinant.

These alternative strategies for production are extremely valuable for vaccines which can not be grown in cell culture or for more complex disease agents (such as parasites). However, there are presently no obvious commercial advantages for these new products as alternatives to vaccines which can be routinely produced by viral propagation in cell cultures.

Monoclonal antibodies may be used for passive immunization which results in short-term disease protection. However, the experimental developments of anti-idiotype antibodies may be significant in introducing a completely novel type of vaccine which may have considerable merits.

From the present plethora of alternative technologies for vaccine development, it is not immediately clear which of these may become the prevailing large-scale production methods in the long-term.

General Reading

Kennedy, R.C., Melnick, J.I. and Dreesman, G.R. (1986). 'Anti-idiotypes and immunity', *Sci. Amer.* **255**, pp. 40–48.

Lerner, R.A. (1983). 'Synthetic vaccines', *Sci. Amer.* **248**, pp. 48–56.

Norrby, E. (1983). 'Viral vaccines: the use of currently available products and future developments', *Arch. Virology* **76**, pp. 163–177.

Parish, H.J. (1965). *A History of Immunization*. New York,Livingstone.

Voller, A. and Friedman, H. (Eds) (1978). *New Trends and Developments in Vaccines*. Lancaster,MTP Press.

Zuckerman, A.J. (1982). 'Developing synthetic vaccines', *Nature* **295**, pp. 98–99.

Specific Reading

Edman, G.C., Hallewell, R.A., Valenzuela, P., Goodman, H.M. and Rutter, W.J. (1981). 'Synthesis of hepatitis B surface and core antigens in *E. coli*', *Nature* **291**, pp. 503–506.

Hopp, T.P. and Woods, K.R. (1981). 'Prediction of antigenic determinants from amino acid sequences', *Proc. Natl. Acad. Sci.* **78**, pp. 3824–3828.

Kennedy, R.C., Eichberg, J.W., Lanford, R.E. and Dreesman, G.R. (1986). 'Anti-idiotypic antibody vaccine for type B viral hepatitis in chimpanzees', *Science* **232**, pp. 220–223.

Mowat, G.N., Garland, A.J. and Spier, R.E. (1978). 'The development of Foot-and-Mouth disease vaccines', *The Veterinary Record* **102**, pp. 190–193.

Salk, J. and Salk, D. (1977). 'Control of influenza and poliomyelitis with killed virus vaccines', *Science* **195**, pp. 834–847.

Smith, G.L., Mackett, M. and Moss, B. (1983). 'Infectious vaccinia virus recombinants that express hepatitis B virus surface antigen', *Nature* **302**, pp. 490–495.

Valenzuela, P., Medina, A., Rutter, W.J., Ammerer, G. and Hall, B.D. (1982). 'Synthesis and assembly of hepatitis B virus surface antigen particles in yeast', *Nature* **298**, pp. 347–350.

Chapter 5

Interferon

Introduction

During the 1930s several observations were made that the infection of an animal by a virus seemed to protect against later infection by a second virus of a different variety. This phenomenon could not be easily explained by the generation of antibodies which are specific to one virus type. This protection against viral infection was later shown in laboratory experiments by Isaacs and Lindenmann in 1957. They showed that chick cells grown in culture could be made resistant to influenza infection by prior exposure to heated non-viable influenza virus. Also, if the fluid bathing the surviving cells was added to other cells, the recipients became resistant to infection from a variety of viruses. They claimed that the fluid contained a compound that interfered with viral propagation and so derived the name *interferon*. They discovered that although this fluid from chick cell cultures could protect other chick cells from various viral infections, this protection did not extend to cells derived from other species. In other words, the interferon was cell specific but not virus specific.

This non-specific anti-viral activity of interferon was compared to the antibacterial properties of the antibiotics and the therapeutic potential of such a compound was immediately recognized. The prospects for the use of interferon for the treatment of viral infection led Gresser and others in the late 1960s to investigate its potential for cancer therapy. This was based on the belief of the viral origin of human cancers. Gresser showed that the growth of tumour cells injected into mice could be retarded by treatment with large doses of interferon. Similar growth inhibition of tumour cells was shown in culture. The results of these experiments were initially greeted with some surprise because of the known non-viral origin of some of the tumours tested. However, this was the first evidence of the property of interferon to retard cell growth and indicated its potential in cancer therapy.

The prospect of interferon being used as an anti-cancer and anti-viral agent was greeted with considerable enthusiasm — particularly by the popular press. However, many of the claims of the properties and therapeutic potential of interferon were made prematurely. The value of such a compound can only be assessed through clinical trials which require several years to perform.

The major problem of initiating clinical trials is being able to obtain sufficient quantities of the material. Many biotechnology companies have been established on the basis of the clinical potential of interferon. They have addressed the problems of supply and purification and these are slowly being solved. However, the assessment of the full therapeutic potential of interferon through clinical trials still requires a lot more work and it may still take several years before all the questions are answered.

Characterization of Interferon

The term 'interferon' refers to inducible secretory proteins produced by eucaryotic cells in response to viral and other stimuli. The complexity of studies on interferon increased with the realization that there exists not just one but a family of related compounds. The original classification divides the interferons into three types related to the originating cells. These are the leucocyte, fibroblast and T-lymphocyte (or immune) interferons. However, this classification was found unsatisfactory because each cell type may produce several types of interferon. The later classification, which is now used, divides the interferons into α, β and γ — which are the dominant interferons derived from the three cell types — leucocytes, fibroblasts and T-lymphocytes (Table 5.1). These three types (IFN-α, IFN-β and IFN-γ) can be clearly distinguished by specific antibodies which do not cross react.

The interferons are inducible proteins and can be further classified according to the nature of the inducing agent. Type 1 (which includes IFN-α and IFN-β) is induced by viruses whereas Type 2 (IFN-γ) is induced by mitogens. A further complexity arises in the discovery of at least 13 independent sub-types of IFN-α which are all products of independent genes although they are sufficiently homogeneous to be indistinguishable by antibody recognition.

Table 5.1 Interferon classification

Type	*Main cell producers*	*Inducers*
α	Leucocytes	virus
β	Fibroblasts	virus or ds RNA
γ	T-lymphocytes	mitogens or antigens

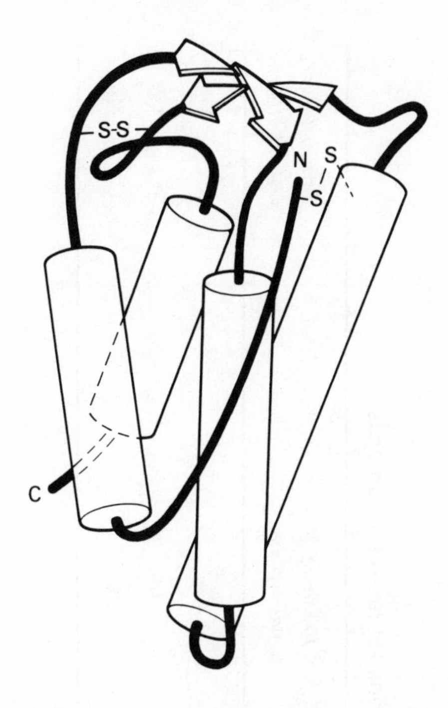

Fig. 5.1 The proposed 3-dimensional structure of interferon.
The figure is a schematic representation of a predicted tertiary structure of interferon based on a consensus of amino and sequences. The components shown are: 4 α helices (cylinders), 2 β-strands (arrows), two disulphide bridges and two peptide termini — C & N (from Sternberg and Cohen, 1982).

Structure

The amino acid sequences of human interferons have been determined by molecular cloning. The IFN-αs consist of 165 or 166 amino acid residues with each sub-type differing between 8 to 29 residues. Two disulphide bridges are known to be present and these serve to maintain the tertiary structure of the molecule. Although IFN-β is distinguished from the IFN-αs by antibody recognition, there are structural similarities. IFN-β consists of 166 amino acid residues, 38 of these being identical in all IFN-αs. IFN-γ can be clearly distinguished from the other interferons by dissimilarities of structure and properties. The protein chain consists of 146 amino acid residues and it is more easily denatured than the other interferons. Although the tertiary structure has not yet been analysed by X-ray crystallography; Fig. 5.1 shows a probable structure predicted from a consensus amino acid sequence of IFN-αs and IFN-β.

Interferons synthesized by mammalian cells initially have a signal peptide consisting of about 20 amino acid residues which is cleaved before the mature protein is secreted. The released interferons may be glycosylated which involves the addition of a carbohydrate group to the protein after translation. The size and

Table 5.2 Properties of interferons found in human cells

Type	*Number of sub-types*	*Number of introns*	*Length of mature protein*	*Length of signal protein*	*Glycosylated*	*Intramolecular cysteine bridges*
α	13	0	165,166	23	No	Yes
β	1	0	166	21	Yes	Yes
γ	1	3	146	20	Yes	No

nature of the carbohydrate, which is typically 18% of the total molecular weight for IFN-β, is variable and this has caused difficulty in characterizing the interferon molecules. IFN-β and IFN-γ contain 1 and 2 glycosylation sites respectively. The IFN-αs are not normally glycosylated although a carbohydrate content for some of the sub-types has been reported. Some of the characteristics of the three main types of interferon are listed in Table 5.2.

Genetics

A multiplicity of genes has been found for the interferon family. The interferons are synthesized in response to specific inducers which cause derepression. The major groups of inducers, which have little in common structurally, are shown in Table 5.1.

At least 13 independent genes for interferon-α have been located on chromosome 9 in human cells. These genes which do not contain introns produce independent protein products which have 90% sequence homology between them. Also, 6 pseudogenes of IFN-αs have been discovered. These do not produce proteins but are characterized by sequences of nucleotides which relate very closely to the functional IFN-α genes and can be identified by hybridization with IFN-α mRNA probes.

Only one interferon-β gene has been isolated by hybridization studies in human cells. It is similar to the interferon-α family of genes in that it does not have an intron and it is also located on chromosome 9. However, evidence exists for the presence of 2 distinct mRNAs for IFN-β which do not cross hybridize. The relationship between these mRNAs and the genetic structure is currently unknown. The interferon genes do vary between species. For example, 3 IFN-β genes have been isolated in bovine cells.

The single IFN-γ gene that has been isolated has little homology with the other IFN genes. It has 3 introns and is located on chromosome 12 in human cells.

Physiological Effects of Interferon

Interferon-α and -β are released naturally by cells *in vivo* in response to viral infection. The level of induction depends upon the state of the cells and is greater in non-dividing cells. However, the specific cellular production is extremely low — 1–10 molecules per cell — and this results in a concentration in the extracellular fluid of 10^{-14} to 10^{-15}M. The effect of this concentration of interferon is to induce an anti-viral state in neighbouring non-infected cells in which viral replication is inhibited. The effect is thought to be an essential component of the normal recovery process from a virus infection (Fig. 5.2).

The nature of this anti-viral state has been the subject of extensive investigation but the molecular mechanism is still not absolutely clear. The interferon does not

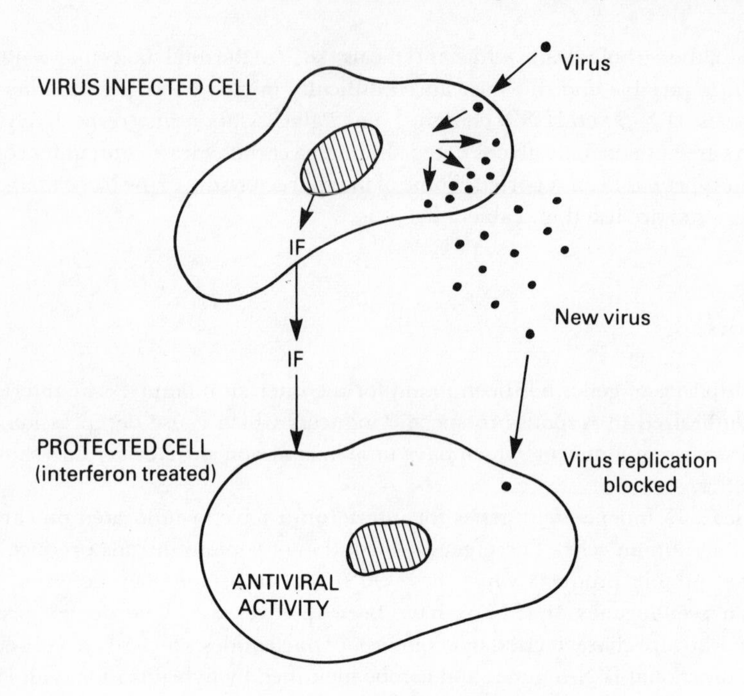

Fig. 5.2 Induction and effect of interferon.

penetrate the cell but attaches to membrane receptors of which 10^3 per cell have been detected for IFN-α and β. Some observed anti-viral effects related to viral attachment, penetration and uncoating may result from changes in the cell membrane stimulated by interferon. Other targets of anti-viral activity include transcription and protein translation. The mechanism of these inhibitory effects are again not fully understood but two enzymes have been implicated. One is the enzyme — 2′5′ oligoadenylate synthetase — which can synthesize a series of oligo-A polymers which activate an endogenous endonuclease capable of degrading RNA. A second enzyme is a protein kinase (p67K) which phosphorylates the eucaryotic initiation factor. The stimulation of the activity of this enzyme by interferon is associated with the inhibition of protein synthesis.

The anti-proliferative effects of interferons which reduce the growth of tumour cells are independent of the anti-viral activities. Although the molecular events of these activities are not fully understood, it was been shown that the protein products of the oncogenes of tumour cells are reduced by interferon treatment.

Interferon-γ can be clearly distinguished from the other interferons. It is one of a number of lymphokines secreted by T-lymphocytes as part of an immune response to both viral or non-viral antigens. The mechanism of stimulation of endogenous synthesis of IFN-γ is complex. IFN-γ would seem to be part of a cascade of lymphokines and its synthesis is dependent upon the secretion of other components including interleukin-1 and interleukin-2 from associated cells in the immune system. The physiological effects of IFN-γ include the anti-viral and anti-proliferative properties which are shared with IFN-α and β but also include the stimulation of an immunological response. This can be partly related to the stimulation of the activity of Natural Killer (NK) cells which are a sub-population of lymphocytes capable of resisting tumour growth. The anti-tumour cell effects of IFN-γ are greater than those of the other interferons and consequently this interferon type has been thought to hold the best possibilities for therapeutic use.

Production of IFN-α from Leucocytes

Much of the early work of developing a process for the large scale production of interferon was conducted in Finland by Cantell. His method of IFN-α production is based on extraction from medium containing human leucocytes which have been induced to produce interferon by exposure to a virus. The leucocytes are obtained by fractionation of large volumes of fresh blood obtained from human donors. A 'buffy coat' preparation is made by precipitation of white blood cells with ammonium sulphate. The precipitate of red blood cell-free leucocytes is then incubated for about 20 h with a virus to induce interferon production. The paramyxoviridae family of viruses has been found suitable for such induction — notably Newcastle disease virus and Sendai virus. In the latter case it has been found advantageous to prime the cells with some interferon before addition of the virus. The treated cells release interferon into the medium. When this is complete, the medium can be isolated and adjusted to a low pH in order to destroy the activity of the virus. This supernatant is a crude mixture from which interferon can be further purified.

The best yields obtained by this process are in the order of 1 μg of purified interferon (10^7 units) per litre of blood, which is less than the quantity of interferon typically required for one daily dose per patient. Such a yield poses real limitations for the development of this process which requires the handling of large volumes of material. The leucocytes must be obtained from fractions of fresh blood and so the interferon induction must be performed soon after blood donation. Units of blood must be pooled from several hundred donors, a procedure which runs the risk of unacceptability through viral contamination (particularly in relation to hepatitis and AIDS).

At the Blood Transfusion Centre in Helsinki this extraction process has been developed for yields of up to 10^{11} units of interferon per year. This has been sufficient for conducting clinical trials and many ideas concerned with the clinical application of IFN-α have been pioneered from this process. However, the potential of this process for scale-up to greater production is limited.

Production of Interferon from Lymphoblastoid Cells

The continuous growth of human cells in culture can solve some of the problems associated with the need for a constant supply of cells from donors. Such a cell line

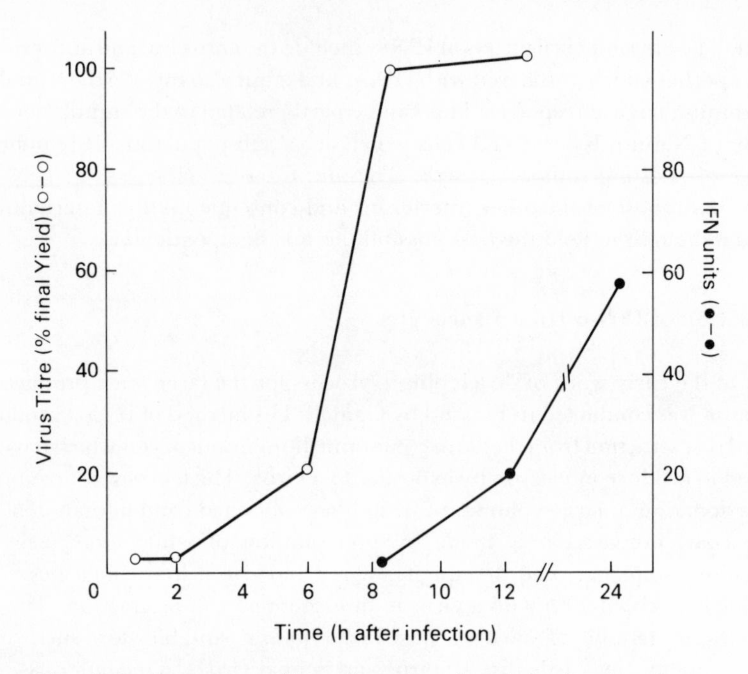

Fig. 5.3 Induction of interferon by virus infection in culture.
(From Friedman, 1981)

suitable for interferon induction has been established from a patient suffering from Burkitt's lymphoma. The resulting 'Namalva' cells are a B-type lymphoblastoid cell line with a near diploid chromosomal complement with the capability of continuous cell division and growth in culture. They are non-anchorage-dependent and therefore can be grown in suspension by the fermentation techniques traditionally used in microbial culture.

Namalva cells have been particularly selected as good producers of interferon. IFN-α is the predominant interferon type induced after about 10 hours incubation with Sendai virus. Figure 5.3 shows typical kinetics of interferon induction in which viral proliferation precedes a period of interferon production by the cells. The interferon is released into the culture medium from which it can be extracted after first deactivating the inducing virus.

The yields obtainable from such a process are 10^8 units of interferon per litre of culture. The method of suspension cell culture is amenable to scale-up and this process is being operated by Wellcome in the UK at a scale approaching 10^4 litres from which gramme quantities of interferon may be obtained. Namalva cells grow well in suspension culture and the cells can be adapted to low cost serum-free media formulations which eases the difficulty of purification of the final product.

The major concern associated with the process is in assessing the safety of using a transformed cell line for the production of a compound intended for therapeutic use in humans. The Namalva cell line has been well characterized genetically. There are 13 characterized chromosomal abnormalities and about one half of the genetic complement of the Epstein Barr virus has been incorporated into the cell's DNA. Although no virus is normally propagated or released by the cells, the original viral incorporation is likely to be associated with the continuous growth characteristics of these cells. It is therefore important to assess the risk of transmitting tumorigenic agents with the use of interferon in clinical treatment.

No DNA associated with the Namalva cells should be allowed to contaminate the final purified interferon product. At first, this could not be guaranteed because of difficulties in purification. However, improved purification techniques and sensitive assays for DNA can now ensure negligible contamination. Thus, despite these concerns it is anticipated that in the near future the risk/benefit ratio of the interferon produced by this process will be sufficient to allow full approval for use in human disease treatment.

Production of Interferon from Human Diploid Fibroblasts

Interferon is synthesized in such small amounts by mammalian cells that to consider an industrial scale production process, large quantities of cells must be handled or the productivity per cell must be increased. The latter approach has been found successful for IFN-β production from human diploid fibroblasts by the process of 'superinduction'.

Human diploid fibroblasts are cells of finite life span with a normal genetic complement derived from surgical biopsies such as male foreskins. These cells are generally maintained up to 30–35 cell passages for interferon production. The cells are anchorage-dependent and therefore require multiple roller bottles or the preferred microcarrier cultures for large-scale production. Interferon synthesis, which is predominantly IFN-β, can be initiated by a variety of non-viral inducers, the best of these found to be a synthetic homopolymer of double stranded RNA — polyriboinosinic:polyribocytidylic acid (poly-IC).

In order to enhance the level of interferon productivity per cell by several orders of magnitude, a regime of antibiotic treatment was developed by Havell and Vilcek in 1972. This is the superinduction process which involves the use of an inducer and two antibiotics. The poly-IC, cycloheximide and actinomycin D are added sequentially to confluent fibroblast cultures by a series of media changes as shown in Fig. 5.4. The rationale behind the superinduction process is to decrease the cellular concentration of an inducible protein repressor which normally causes the breakdown of interferon mRNA. Cycloheximide is a reversible inhibitor of protein synthesis and prevents the formation of the repressor (as well as interferon) in the 5 h period in which it is added to the cell culture. However, during this period the synthesis of interferon mRNA which has been induced by poly-IC proceeds normally. At the end of this period the IFN mRNA has accumulated to a sufficiently high level. Interferon synthesis is then initiated by the removal of cycloheximide. However, the addition of actinomycin D at this stage inhibits transcription which particularly affects the synthesis of repressor mRNA.

The absence of the repressor increases the half-life of IFN mRNA by a factor of

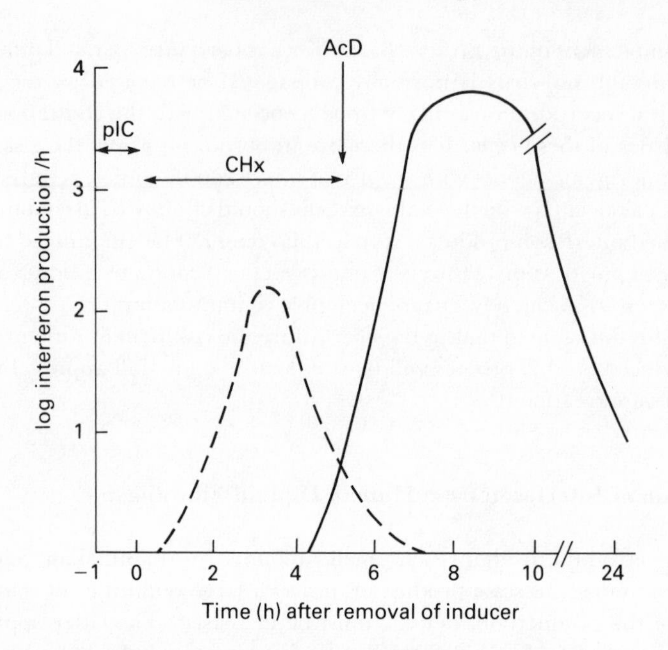

Fig. 5.4 Superinduction of IFN-β in human diploid fibroblasts. Interferon production is shown after induction with poly I-C and with antibiotic treatment (superinduction) (——) or without antibiotic treatment (----).
(From Friedman, 1981)

$\times 3$ to $\times 4$. The overall effect of this is to delay the induction of interferon in these cells and to increase the total production of interferon by a factor of $\times 20$ to $\times 100$.

The use of the superinduction process to increase the cellular production of interferon has allowed the development of an industrial production process based on cells which are relatively difficult to handle in large quantities because of their anchorage dependence. Scale-up of these cultures on microcarriers is preferred because of the ease of handling homogeneous cultures in which changes of media are simple. The cells have an advantage in that they are derived from tissue with a normal chromosomal complement. This eases the problem of acceptance of the product by drug regulatory authorities. At the present time the process of IFN-β production from human diploid fibroblasts is operational at a scale of several hundred litres by various companies who are preparing for clinical trials.

Interferon from Recombinant Bacteria

The ability to obtain large quantities of interferon has been considerably advanced by recombinant DNA technology which allows the insertion of a human interferon gene into a bacterium. In 1980, Goeddel and his colleagues at Genentech reported the production of human IFN-α for the first time from a transformed *E. coli* culture. Subsequently, these techniques have been applied to the production of other types of interferon. Although various strategies have been employed, the methods used by Goeddel will be described here as an example of the construction of a bacterium capable of human interferon synthesis (Fig. 5.5).

1. A human myeloblastoid cell line was chosen as a source of IFN-α mRNA. This cell line was established from myeloblasts isolated by biopsy from a man suffering from leukaemia. The cultured cells were induced to synthesize interferon by infection with Sendai virus
2. The mRNA in these virally-induced cells was isolated by affinity chromatography by selective attachment to particles containing oligo-dT. In eucaryotic cells messenger RNA can be distinguished from other RNAs by the presence of a chain of adenosine units at the 3′ end — the poly-A tail. The tail allows the hybridization of the poly-A containing RNA to the oligo-dT particles while the unwanted components of the cell lysate are washed through the column. A change in pH of the eluting solution disrupts the hybridized complementary bases and a poly-A RNA rich fraction can be collected. At this stage the IFN mRNA constitutes less than 1% of the total isolated poly-A RNA
3. To enrich further for interferon mRNA, the solution was fractionated by size on sucrose density gradients. Fractions of 12S were selected. This procedure particularly selects for the poly-A — containing molecules that are complete mRNA molecules rather than fragments
4. At this stage the interferon mRNA content of the isolated fraction was tested by micro-injection into Xenopus oocytes. These cells are chosen because of their physical size which makes micro-injection possible. Messenger RNA from any eucaryotic source can be tested in these cells for protein production. By this technique, the poly-A rich 12S fraction was shown to be capable of interferon synthesis
5. The next stage was to convert the mRNA into complementary double stranded DNA (cDNA) by the use of a reverse transcriptase enzyme. Following this conversion, cDNA of the appropriate length was recovered by electrophoresis. This pure cDNA has a nucleotide sequence corresponding to the human gene but without the promoter-operator regions and those sequences that may be removed after mRNA transcription
6. The cDNA was prepared for insertion into a plasmid by homopolymer tailing. This involves the enzymatic addition of a series of deoxycytosine (dC) nucleotides on each of the 3′ ends of the cDNA
7. The commonly used circular plasmid — pBR322 — was opened at the single PstI restriction site which is present in the ampicillin resistant gene of the plasmid. The 3″ ends of this opened plasmid was tailed with a deoxyguanosine (dG) sequence
8. The tailed sequences of DNA prepared in [6] and [7] were mixed and a recombinant DNA was allowed to form by annealing which was promoted by the complementary nature of the tailed sequences of dC and dG

(1)

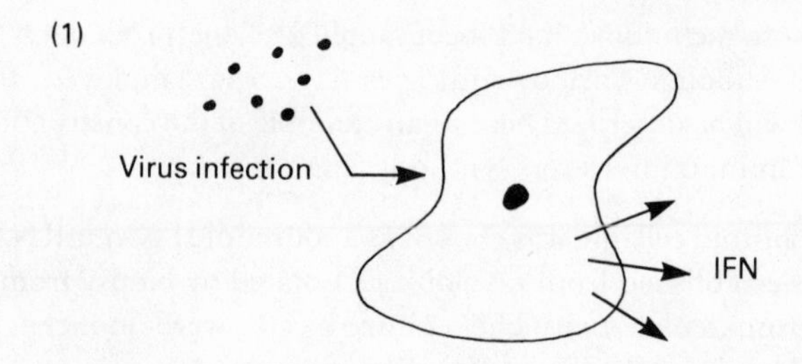

Myeloblastoid cell line induced to synthesize interferon

(2)

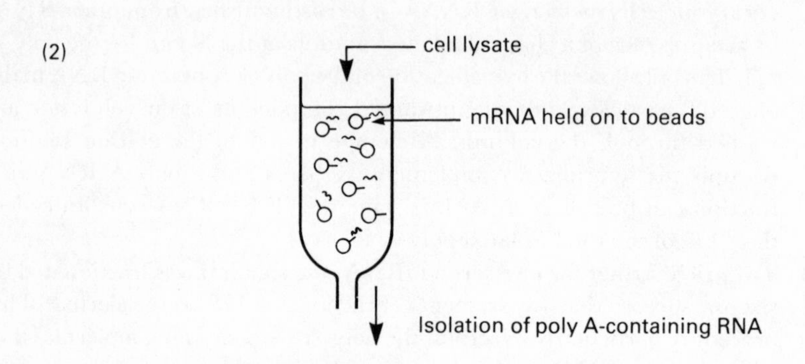

(3)

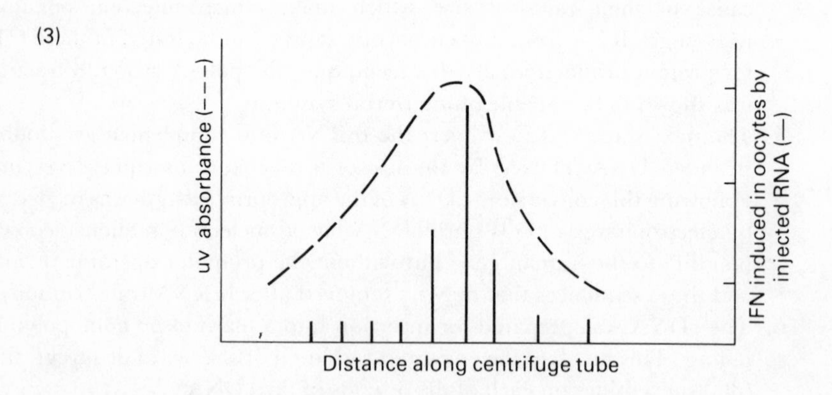

Size fractionation of RNA by sucrose density centrifugation

Fig. 5.5 Isolation of interferon mRNA.

9. The DNA mixture, only a proportion of which was recombinant, was used to transform a culture of *E. coli*
10. Tetracycline-resistant and ampicillin-sensitive clones of *E. coli* were selected. This combination of reactions to the antibiotics could only be indicative of the presence of a recombinant plasmid. The procedure in which the ampicillin resistant gene is interrupted is known as insertional inactivation. Although these selected clones possess the interferon gene, at this stage they are not capable of gene expression, i.e. synthesis of interferon. The clones were grown in order to produce large quantities of the interferon gene
11. These clones are not capable of interferon synthesis because of the absence of a promoter region. The promoter is a sequence of nucleotides upstream of any gene and differs in structure between procaryotic and eucaryotic cells. For gene products to be formed in *E. coli*, a procaryotic-type promoter is required.

    A suitable promoter was obtained from part of the well characterized *E. coli* tryptophan operon. This consists of the early part of the operon but without the sequences required to form a functional protein. This promoter was inserted into the recombinant plasmid which was now available in reasonable quantities by previous molecular cloning. Insertion of this promoter just upstream of the interferon gene produces a plasmid capable of initiating interferon synthesis in *E. coli*. The selected *E. coli* clones were capable of producing 2.5×10^8 units of interferon per litre of culture (specific activity $= 4 \times 10^8$ units mg^{-1})
12. The efficacy of the interferon produced by the *E. coli* transformed by this functional plasmid was then tested. The antiviral activity of the genetically engineered interferon was shown by its ability to protect squirrel monkeys against the lethal effects of the encephalomyocarditis (EMC) virus.

E. coli cultures have now been transformed by recombinant plasmids constructed for the synthesis of most of the known human interferon types. The characterization of the interferon genes has enabled the construction of probes which can be used to isolate mRNAs of individual interferon sub-types. One of the most successful of these is a chemically synthesized oligonucleotide consisting of 15 nucleotides based on a highly conserved region of IFN-α. Such isolates have enabled the construction of bacterial clones capable of large-scale production of individual interferon sub-types.

However, there are some disadvantages of interferon production from recombinant bacteria. The product is not glycosylated as it is in mammalian cells. Although such non-glycosylated interferon has been shown to be biologically active, its tertiary structure may be less stable. This may lead to a lower specific activity *in vitro* and a faster rate of breakdown *in vivo*. Another difference is that the recombinant bacterial interferon is a single homogeneous molecular species as opposed to the heterogeneous mixture of types produced from mammalian cells. This can be advantageous for molecular characterization but the specific mixture of types secreted by selected cell lines may have some therapeutic advantages.

Production of IFN-γ by Cloned Genes in Eucaryotes

Human IFN-γ was first described by Wheelock in 1965 who reported its induction from human white blood cells by phytohaemagglutinin. Interest in this interferon type grew with the discovery of its unique properties of immunoregulation and anti-tumour activity. Its synthesis by T-lymphocytes can be stimulated by a range of antigens or mitogens. The most popular of these has been the bacterial product — staphylococcal enterotoxin A. This is a powerful T-cell mitogen which stimulates the extracellular release of IFN-γ by human peripheral blood lymphocytes obtained from volunteers. By this method, induction takes 3–5 days with maximum levels of 10^7 units of interferon produced from 4 l of blood.

The problems of obtaining the large quantities of product required for clinical use are similar to those indicated earlier for IFN-α produced from leucocytes by Cantell's method. These difficulties have led several research groups to consider the insertion of cloned IFN-γ genes into eucaryotes. The techniques of gene manipulation are not as advanced in eucaryotes as compared to procaryotes but there are a number of potential advantages (see Chapter 3). In 1983, Fiers reported the expression of human IFN-γ in Chinese hamster ovary (CHO) cells and some of the techniques used are as follows:

(1) A complementary DNA (cDNA) was prepared from mRNA extracted from induced human T-lymphocytes. The methods used were similar to those used in the preparation of cDNA to IFN-α as indicated earlier
(2) The isolated cDNA was inserted into a previously constructed plasmid which contained the gene promoter for the simian virus — SV40. The purpose of this was to allow the expression of the IFN-γ gene to be under the control of this promoter which is constitutive. In other words, the production of the interferon would no longer depend upon the addition of inducers
(3) The CHO cells chosen were deficient in the gene for the enzyme — dihydrofolate reductase (DHFR). This enzyme is necessary for the cellular synthesis of tetrahydrofolate which is a co-enzyme required for the synthesis of purines and pyrimidines (see Chapter 3). Therefore CHO cells which are DHFR$^-$ can only be grown in a media containing ready made purines and pyrimidines usually supplied as ribo- and deoxyribo-nucleosides
(4) These mutant CHO cells were transfected by two plasmids simultaneously. One plasmid contained the IFN-γ with the SV40 promoter and the second contained the gene for DHFR
(5) The cells were incubated in a selective medium which was lacking in ribo- or deoxyribo-nucleosides. By such selection only DHFR$^-$ cells survived
(6) The selected cells were then grown in progressively increased concentra-

tions of methotrexate. This compound is an inhibitor of the enzyme DHFR but at certain concentrations, cells may respond by amplifying their DHFR gene content and the metabolic block can be overcome by an increased concentration of the enzyme. The cells that are co-transfected with other plasmid genes will also show co-amplification of these genes. Therefore, from such an amplification procedure cells can be selected that have several copies of the plasmid containing the IFN-γ gene. Such selected cells that have been grown in culture have been shown to be constitutive producers of IFN-γ at high specific production rates.

Subsequently, it has been reported that CHO cells transfected by plasmids containing the human IFN-γ gene derived by genomic DNA fragmentation show even higher specific production rates. The presence of introns may facilitate expression in these transfected genes. Such genetically manipulated eucaryotic cell lines can produce up to $\times$ 500 more IFN-γ per cell than cultured T-lymphocytes.

Other research groups have attempted the transfection of lower eucaryotes such as yeast for interferon gene expression. Yeast is an attractive host for protein production for a number of reasons. In common with other eucaryotic cells, it has the advantage of being able to process proteins after translation and secrete the products into the medium. It has the further advantage of withstanding high hydrostatic pressure and it does not lyse after death. The growth medium is particularly simple and this aids protein purification. A number of plasmids have been constructed to direct human interferon synthesis in yeast. These require a human signal sequence, the presence of which is essential to allow appropriate post-translational modification and protein secretion.

Novel Interferons

The techniques of molecular cloning can be used to create novel individual molecular species of interferon. For example, hybrid interferons from two distinct naturally occurring types or sub-types can be created by fragmenting gene isolates by using a convenient common restriction site. A hybrid gene can then be created by ligating the terminal fragment of one gene with the opposite terminal of a second gene. The value of this work lies in attempts to establish the relationship between structure and biological properties of the molecule. It would be desirable if this could lead to the creation of a novel interferon which possessed enhanced therapeutic properties but with limited undesirable side effects. This approach of producing interferon analogues is being persued by Stebbing and others at Amgen in the USA. They have reported improved antiviral activity with a genetically constructed IFN-α based on a consensus structure of all the 13 known sub-types.

Chemical Synthesis of an Interferon Gene

In 1981, Edge developed the phosphotriester method for polynucleotide synthesis

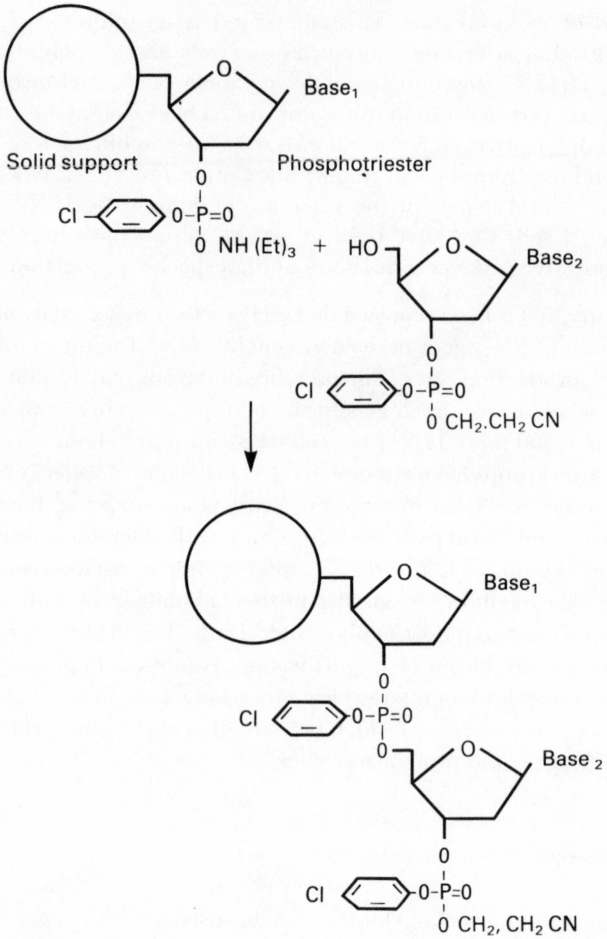

Solid support
Base$_1$
Phosphotriester
Cl
$O-P=O$
$O\ NH(Et)_3$ + HO
Base$_2$
$O\ CH_2.CH_2\ CN$
Base$_1$
Base 2
$O\ CH_2, CH_2\ CN$

Fig. 5.6 Sequential addition of protected nucleotides on solid phase support material.

which was applied to the construction of an IFN-α gene. This was a remarkable achievement because of the size of the gene — 517 base pairs.

The method of solid phase synthesis is based on the sequential addition of chemically protected nucleotides to a column of solid phase support material which traps the growing oligonucleotides (Fig. 5.6). The free 5′ hydroxyl group of the incoming nucleotide is allowed to react with a fully masked 3′ phosphotriester group of a nucleotide attached to the solid phase. By careful addition of protective and reactive groups to the nucleotides, the desired 3′ 5′ phosphoester bridge can be formed and thus extending the length of the oligonucleotide. After a series of such condensation reactions the blocking groups can be removed by hydrolysis.

This method can be used for the synthesis of an oligonucleotide of up to 20 nucleotide residues which can then be isolated by HPLC. In Edge's work, 66 such individual oligodeoxyribonucleotides were prepared before being linked together enzymatically to form the complete interferon gene. This gene can be used for interferon synthesis in a transformed *E. coli* culture.

These procedures of chemical synthesis of the gene have particular value in creating new forms of interferon by deliberate modifications of the synthesized nucleotide sequence. This can allow the determination of the sites of biological activity or the creation of novel interferons with enhanced therapeutic activity.

Assays for Interferon

Progress in the development of large-scale processes for the production of interferon has been hampered by difficulties in assaying the compound. The original assays developed were based on *in vitro* tests of antiviral activity which were sensitive but gave highly variable results. Comparison of results between laboratories was eased by the standardization of the assay by the World Health Organisation (WHO). The production of a standard reference preparation of interferon suitable for this assay allowed all other preparations to be calibrated in International Units.

Antiviral Assays

Several interferon assays have been based on the inhibition of the cytopathic effect of viruses on cultured cells. These include direct microscopic examination of infected cells and the inhibition of the uptake of vital dyes. Alternatively, interferon-induced reduction in virus yields can be measured in a variety of ways including assays for activities of haemagglutinin, neuramidase, reverse transcriptase, viral RNA synthesis or plaque formation. A typical assay based upon reduction of viral plaque formation involves the following.

(1) Cultured cells are treated for 14h with serial dilutions of an interferon preparation
(2) The cells are washed and infected with virus at a suitable virus/cell ratio
(3) After 8h (which is the end of rapid viral replication) the cell cultures are frozen
(4) The plaque forming units (p.f.u.) of the thawed media samples are determined on cell monolayers
(5) The interferon activity is then expressed as the inhibition of virus growth determined from the difference between the virus yields (p.f.u's) in the treated samples and the untreated controls

The assay of biological activity of interferon in such a test is particularly sensitive (~0.5pg IFN). Results can be compared with the International Reference Preparation as calibrated by WHO and expressed in International Units of activity. However, although this standardization reduces the variability between independent assays and allows comparison between different laboratories, there are still some problems. The antiviral activity of an interferon preparation may not be correlated with other activities, e.g. anti-tumour or immunoregulatory. Also, comparison of individual interferon sub-types and the mixtures generally used as references may not be valid over a range of assays.

A further drawback of the antiviral assay for use in developing a production process for interferon is the time it takes to obtain results — typically 3 days. This has encouraged the use of immunoassays which have found wide acceptance following the production of monoclonal antibodies against interferons.

Immunoassays

The major advantage of an immunoassay is the time it takes to obtain results — hours rather than days in the antiviral assays. This is particularly advantageous for the development of processes of production and purification. Monoclonal antibodies have been raised to specific interferons and assays have now been designed to distinguish between interferons even of subtypes of very similar structure. These assays can be based on radioimmunoassays (RIA) or on enzyme-linked-immunosorbent assays (ELISA) — see Chapter 6.

If the molecular structures of the test and reference interferons are identical, comparisons between immunoassays and antiviral assays are possible. Such comparisons between assays are particularly useful in the determination of specific molecular activity i.e. biological activity, per unit weight.

Purification of Interferons

The small quantities of interferon released by cultured cells has resulted in problems concerned with extraction and obtaining pure products. In the early days of the development of Cantell's method of interferon production from leucocytes, interferon extracts of less than 1% purity were tested in clinical trials. The low purity of such samples resulted in great difficulties in the interpretation of results and undesirable side-effects were attributed to impurities.

Purification improved considerably with the development of monoclonal antibodies against interferon. The high specificity of the antibodies enables the isolation of interferon from mixtures of unwanted macromolecules. The general procedure is to bind the antibody to inert particles such as Sepharose which can be held in a column. Control of the pH of the buffer running through the column can cause the binding and release of the interferon to be purified. Such immunoaffinity columns are re-usable and a relatively small column is suitable for extraction from large quantities of crude material. Data from Hoffman-La Roche in their large-scale cultures of genetically engineered *E. coli* show that a 17 ml column is suitable for a $\times 10^3$ purification step from a 1 kg bacterial extract (Table 5.3). Figure 5.7 shows a typical 3-step purification procedures used for IFN-α where the second step of immunoaffinity chromatography produces a dramatic increase in specific activity and at a high percentage recovery. The final interferon extract from this process is sufficiently pure that it can be crystallized. There are a

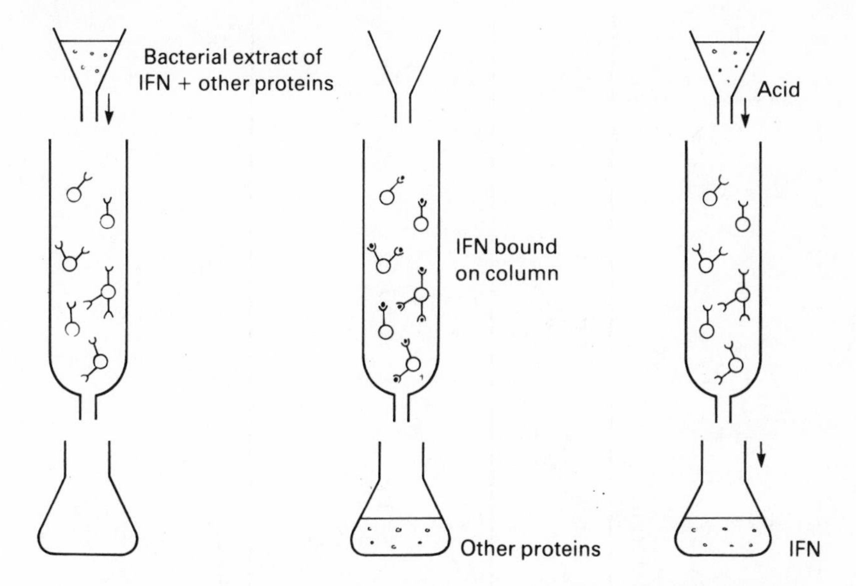

Fig. 5.7 Purification of interferon by an immuno affinity column.

number of advantages in producing interferon at such purity. It is much easier to characterize the molecular structure and biological activity of the interferon. It also increases the acceptability of the use of the compound in human therapy.

Therapeutic Potential of Interferon

The prospects for the use of interferon as an anti-viral therapeutic agent was anticipated from its original discovery. However, it is now considered that its effective use as an anti-viral agent may be limited to those clinical situations in which the endogenous synthesis is low or where it can be administered before viral infection. Low endogenous synthesis may occur in acute viral infections that are protracted or in immunocompromised patients. Some successes have been found in the treatment of hepatitis B, herpes virus and respiratory tract virus infections. Also, interferon has been effective in preventing cytomegalovirus infection, which is a common complication of transplant operations. Trials with the common cold suggest that prolonged intranasal treatment with interferon protects against later infection. However the high doses required and the unpleasant side-effects limit its widespread use for this purpose. Some veterinary uses are also possible, for example, in protecting vulnerable cattle against bovine shipping fever.

The discovery of the anti-proliferative effects of interferons immediately suggested their therapeutic use in the treatment of cancers. However, the full realization of these prospects has been thwarted by the complexities of the interferon system and the difficulties in obtaining large quantities of pure material.

Table 5.3 Purification of interferon-α

Step	*Volume (ml)*	*Total protein (mg)*	*Total activity (units)*	*Specific activity (units/mg)*	*Purification factor*	*Recovery %*
$(NH_4)_2SO_4$ fractionation	700	37 100	7.4×10^9	2.0×10^5	1.0	100
Antibody column separation	27	30	7.0×10^9	2.3×10^8	1150	95
CM-cellulose separation	30	20	6.0×10^9	3.0×10^8	1500	81

Note The data was taken from the purification of recombinant IFN-α produced by Hoffman-La Roche.

Nevertheless, some of the clinical trials that have been carried out have shown the value of interferon therapy and that particular interferon types should be used preferentially in the treatment of certain viral diseases or cancers.

Trials of the use of IFN-α extracted from leucocytes by the Cantell process have shown impressive results in selected conditions such as the treatment of hairy cell leukaemias and juvenile laryngeal papilloma. The latter is non-malignant but life-threatening if untreated. Some of the recombinant IFN-α's have been shown to be effective against some melanomas and renal cell cancers which are unresponsive to other therapies. Similarly, IFN-β has shown promise in the treatment of solid tumours. IFN-γ has engendered considerable promise in various clinical situations because it is a powerful stimulator of the immune system as well as having enhanced anti-tumour properties.

Differences in the nature and timing of the effects produced by the various interferon types have suggested the value of using mixtures of interferons in therapeutic treatment. One major drawback in the clinical use of the interferons concerns their side-effects. These include fatigue and flu-like symptoms which are particularly pronounced at the continuous high doses sometimes found necessary. Although originally thought to be associated with impurities, these side-effects have now been shown even with highly purified preparations. One approach that may alleviate this problem and allow the administration of lowered dose levels is to conjugate the interferon to a monoclonal antibody specific for a tumour cell membrane marker. This has been shown to be successful in in vitro studies and may have future clinical potential.

Despite the success of some of the clinical trials, there are still some major questions to be addressed before the interferons can realize their full therapeutic potential. These questions include establishing the optimal interferon types and optimal dose levels to be used in particular clinical conditions. Also, the efficacy of natural vs recombinant interferons needs to be determined. Answers to these and other similar questions regarding clinical usage, clearly affect the methods of large-scale production which will be continued and developed further.

Conclusion

From the initial discovery of interferon in 1957, difficulties in the assessment of its therapeutic potential have been associated with the exceedingly small quantities of the product naturally obtainable from cells. Over the last few years considerable advances have been made in the problems of production. Developments in cell culture techniques and recombinant DNA technology have allowed the routine production of various interferon types from fermenters of capacity above 10^3 litres. The problems of purification have also been tackled. The use of specific monoclonal antibodies allows purification factors of $> \times 10^3$ in a single step and has now enabled the isolation of interferons of high specific activity.

Future developments of the technology of interferon production depend upon the success of the clinical trials which are on-going. The full picture of the therapeutic use of interferons is still likely to take several years to emerge because of the multiple effects associated with a multiplicity of interferon types which may be useful in the treatment of a range of pathological conditions. The full complexity of the clinical possibilities needs to be unravelled.

It is likely that for the near future the variety of production methods described here will continue to be used at least for the duration of the clinical trials. The favoured method of production will eventually depend upon the efficacy of the interferon produced as well as the rates of production. Questions such as the importance of glycosylation and other post-translational events will determine the usefulness of recombinant DNA production methods in bacteria. However, the use of genetically engineered eucaryotic cells holds promise because of the double advantage of high production rates and a fully glycosylated product. Also, production methods involving the induction of cultured cells should not be ignored. Such methods may be boosted by the increased acceptance of the use of transformed cells which are easy to grow.

The relative merits of these methods of production are presently in the balance. The next few years will show which of the various technologies can lead to the production of efficacious interferon at reasonable cost.

Summary

Interferon was discovered in 1957 as a naturally produced protein which interferes with virus infections. Later discoveries revealed that there are a family of similar molecules which have now been classified into three groups — α, β and γ — representing the major types produced by leucocytes, fibroblasts and T-lymphocytes. In the case of human interferon-α, there are at least 13 sub-types, each showing slight amino acid sequence differences.

The interferons include anti-viral, anti-proliferative and immunoregulatory activities — such properties, which have been demonstrated *in vitro*, have generated considerable interest in their clinical potential as therapeutic agents against viral diseases and cancers. This potential has not been fully realized because of the difficulties in producing sufficient quantities of the purified products to run clinical trials.

In the last few years some of the technology concerned with large-scale production and purification has made considerable advances. In particular, improvements in the techniques of the growth and induction of human cells in culture have increased specific production rates. Techniques of genetic manipulation have resulted in the construction of both procaryotic and eucaryotic cells capable of high expression rates from inserted genes. Production of monoclonal antibodies against interferons has allowed the design of faster and more discriminating assays which have helped the development of the large-scale production processes. Further, these monoclonal antibodies can be used to purify interferons to levels previously unattainable.

At the present, purified interferons of various types can be produced in quantity by a multitude of different processes. The clinical trials necessary to assess the efficacy of these interferons are on-going. Future developments in the technology of interferon production now depend upon the results of these trials.

General Reading

Bollon, A.P. (Ed.) (1984). *Recombinant DNA Products: Insulin, Interferon and Growth Hormone.* Florida, CRC Press.

Finter, N.B. (Ed.) (1984). 'Interferon', Vols. 1–4. Barking, Elsevier Applied Science.

Friedman, R.M. (1981). *Interferons* — A Primer. New York. Academic Press.

Kirchner, H. and Schellekens, H. (Eds) (1984). *The Biology of the Interferon System.* Barking, Elsevier Applied Science.

Panem, S. (1984). *The Interferon Crusade.* Washington, Brookings Inst.

Taylor-Papadimitriou, J. (1985). *Interferons: Their Impact in Biology and Medicine.* Oxford, Oxford Univ. Press.

Zoon, K.C., Noguchi, P.D. and Liu, T.-Y. (Eds) (1984). *Interferon: Research, Clinical Application, and Regulatory Consideration.* Barking, Elsevier Applied Science.

Specific Reading

Allen, G. and Fantes, K.H. (1980). 'A family of structural genes for human lymphoblastoid (leukocyte-type) interferon, *Nature* **287**, pp. 408–411.

Derynck, R. *et al.* (1980). 'Expression of human fibroblast interferon gene in Escherichia coli', *Nature* **287**, pp. 193–197.

Edge, M.D. *et al.* (1981). 'Total synthesis of a human leucocyte interferon gene', *Nature* **192**, pp. 736–762.

Goeddel, D.V. *et al.* (1980). 'Human leucocyte interferon produced by *E. coli* is biologically active,' *Nature* **287**, pp. 411–416.

Gray, P.W. *et al.* (1982). 'Expression of human immune interferon cDNA in *Escherichia coli* and monkey cells,' *Nature* **295**, 503.

Hitzeman, R.A. *et al.* (1983). 'Secretion of human interferons by yeast,' *Science* **219**, pp. 620–625.

Isaacs, A. and Lindenmann, J. (1957). 'Virus interference 1. The interferon', *Proc. R. Soc. London* **147**, 258.

Lazar, A. (1983). 'Interferon production by human lymphoblastoid cells,' *Adv. Biotech. Processes* **2**, pp. 179–208.

Moore, M. and Dawson, M.M. (1987) 'Interferons' in *Textbook of Immunopharmacology.* Dale and Foreman Eds), Oxford Blackwell Scientific.

Pestka, S. (1983). 'The purification and manufacture of human interferons,' *Sci. Amer.* **249**, pp. 29–35.

Powledge, T.M. (1984). 'Interferon on trial,' *Biotechnology* **2**, pp. 214–228.

Scahill, S.J. *et al.* (1983). 'Expression and characterization of the product of a human immune interferon cDNA gene in Chinese hamster ovary cells', *pNAS* **80**, pp. 4654–4658.

Shoham, J. (1983). 'Immune interferon production,' *Adv. Biotech. Processes* **2**, pp. 209–269.

Wheelock, E.F. (1965). 'Interferon-like virus inhibitor induced in human leucocytes by phytohemagglutinin,' *Science* **149**, pp. 310–311.

Chapter 6

Monoclonal Antibodies

Introduction

The immunological system of all higher animals is capable of recognizing a foreign compound as an antigen and of producing a complementary antibody which is highly specific in its binding capacity. Antibodies are glycoprotein products of a sub-set of white blood cells — the B-lymphocytes. The population of B-lymphocytes in any animal is capable of producing a range of antibodies, the nature of which depend upon previous exposure to antigens. The diversity of antibodies which can be produced and the genetic variability required for this have been fundamental biological problems which have been debated for decades and are only now gradually being understood.

The usefulness of an antibody outside the living system relates to its ability to bind a particular compound with high specificity. However, any attempt to isolate antibodies from an animal results in a heterogeneous mixture which depends upon the history of antigenic exposure. Although such polyclonal mixtures have been useful experimentally their widespread use has severe limitations because of the variability of each preparation.

This situation changed dramatically in 1975 when Kohler and Milstein from Cambridge described a technique to produce cells capable of continuous secretion of a single type of antibody to a predefined antigen. Since then, this technique has resulted in the development of a technology which is now capable of producing kilogramme quantities of an antibody to any antigen of choice. The availability of such quantities of homogeneous antibody isolates has resulted in an increasing list of useful applications.

Structure and Classification of Antibodies

There are five recognized classes of antibodies — termed immunoglobulin G,A,M,D and E. Each of these classes differs in peptide chain structure and physiological location. Immunoglobulin A (IgA) is the major class found in external secretions such as saliva, tears and mucus. Immunoglobulin M (IgM) is the first class to appear in the blood serum after antigenic exposure whereas IgG eventually becomes the most abundant class in serum and is the most well understood.

Figure 6.1 shows the structure of IgG which has a molecular weight of 150 kilodaltons and consists of 4 peptide chains held together by disulphide bridges. The primary structure of the molecule is variable and this reflects the diversity of binding pattern. In particular, an extremely variable region is found at the N-termini of the 4 chains which coincides with the region capable of antigen binding. This antigen binding region which is unique to a particular antibody is referred to as an idiotype (from Greek meaning 'individual form'). The specificity of binding is determined by the 3-dimensional shape of the N-termini region of a paired heavy and light chain which complements the shape of a portion (epitope) of the antigen. Small antigens may possess one epitope whereas a large antigen may possess several. This means that in some cases several antibodies of differing structure may be generated for one antigen. The nature of antigen-antibody binding is analogous

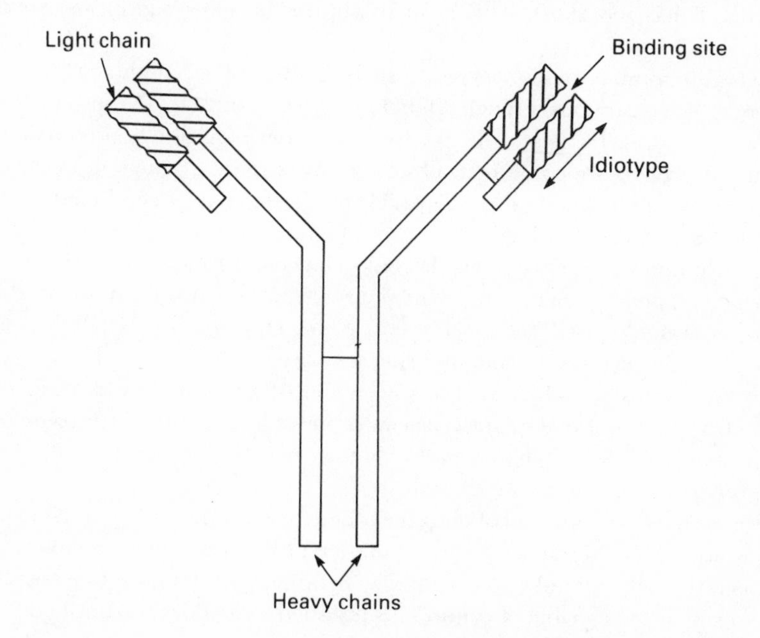

Fig. 6.1 Structure of an immunoglobulin antibody. The variable region and antigen binding sites are shown at the N-termini of the peptide chains.

to that of enzyme-substrate. The interactions are non-covalent and depend upon electrostatic, hydrogen bonding and non-polar interactions to maintain a strong and specific binding.

The mechanism by which a diverse series of proteins may be produced to a seemingly infinite number of antigens has received considerable attention and analysis. There are many domains of the protein structure coded for by genes located at various sites in the genome. The combination of these genes and subsequent splicing during lymphocyte differentiation seems to be responsible for this diversity of protein structure.

Development of Hybridoma Technology

The fundamental background to the development of hybridoma technology is the clonal selection hypothesis of Macfarlane Burnet. This hypothesis states that each mammalian B-lymphocyte is capable of producing only one antibody type with a single specificity. Stimulation by an antigen causes a B-lymphocyte cell having receptors for that particular antigen to expand into a population of plasma cells all secreting antibody of a single type and specificity. The variety of antibodies present in an animal is a reflection of the population of B-lymphocytes — each one of which may synthesize a different antibody. It is possible to select a single lymphocyte capable of synthesizing one antibody type. However, such an isolated cell has too short a life-span in culture to allow significant antibody production.

In certain human malignancies — such as Burkitt's lymphoma and some myelomas — continuous growth of lymphocytes results in overproduction of specific antibodies. In these cells, the two properties of antibody production and continuous growth are combined. On the basis of these findings, it occurred to Kohler and Milstein that these two desirable properties may be combined *in vitro* by fusing two selected cell lines.

The techniques of cell fusion had been developed a decade earlier using various fusion agents, notably inactivated Sendai virus. This has now been superseded by a more commonly used chemical agent — polyethylene glycol (PEG). These agents cause the seemingly random fusion of cells which may grow through subsequent generations usually with some rejection of chromosomal material. The fusion of normal and malignant cells can produce hybrids in which some of the genetic characteristics of the normal cells may be 'immortalized' in the hybrid which is capable of continuous growth.

It was with the background of these ideas and fusion techniques that Kohler and Milstein were able to immortalize a population of B-lymphocytes capable of antibody secretion. By careful selection of the resulting hybridomas it was possible to isolate a single cell capable of continuous production of a desired antibody.

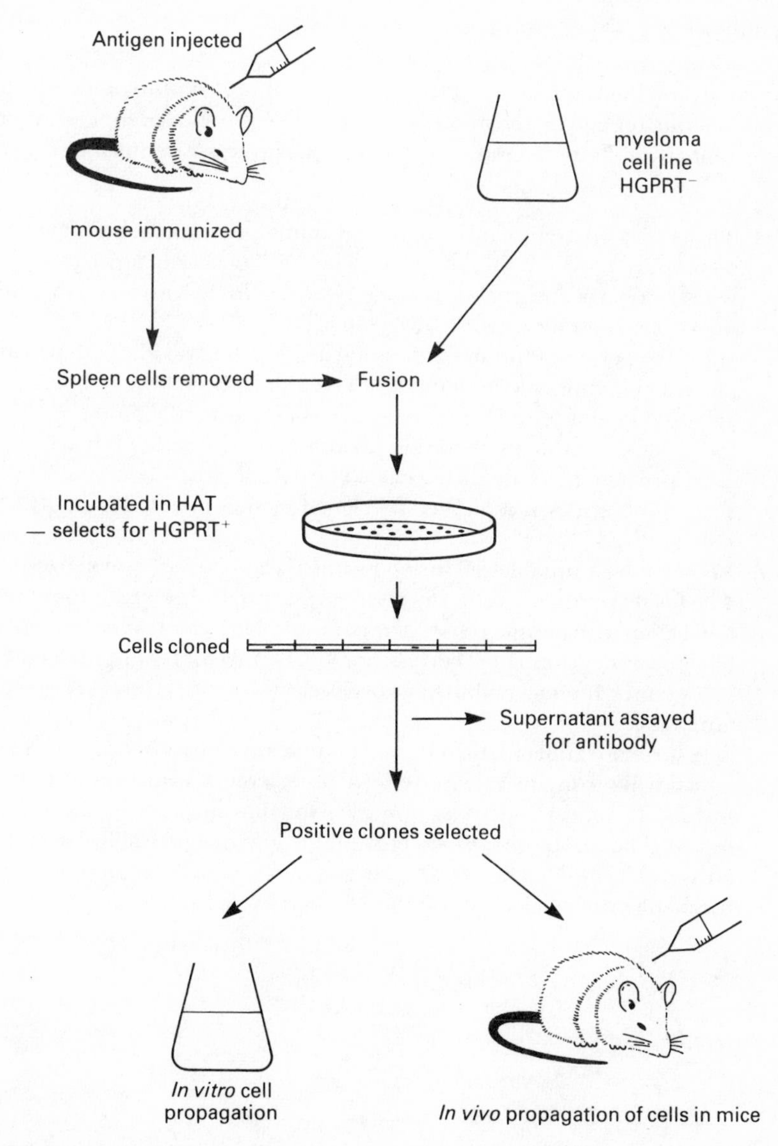

Fig. 6.2 Construction and selection of a monoclonal antibody-secreting hybridoma.

Production of Monoclonal Antibody-Secreting Hybridomas

Monoclonal antibodies can be produced against any compound which is recognised as an antigen by the immune system. The procedures for constructing and selecting an antibody- secreting hybridoma of choice are outlined in Fig. 6.2. The stages involved are as follows:

(1) The chosen antigen is injected into an animal — mice and rats have been commonly used for this purpose. The immunization procedure varies with the type of antigen selected but typically this will involve two or three injections of antigen over a 3 week period

(2) After the period of immunization during which antibodies are produced against the antigen, the animal is killed and the spleen removed. The spleen contains developing cells associated with the immune system including the antibody-secreting plasma cells. The spleen is macerated to a suspension of individual cells. The lymphocytes contained in this suspension are then separated by centrifugation in a dense fluid such as Ficoll

(3) Meanwhile, a suitable cell fusion partner is chosen and grown in culture. The fusion partner is a transformed cell line capable of continuous growth and fusion with plasma cells. A range of suitable rodent myeloma cell lines have been developed for such fusions. The myelomas are derived from malignant lymphocytes but have been selected for their inability to secrete antibodies.

A further characteristic required of these myelomas is a genetic marker which will distinguish them from the hybridomas which result from cell fusion. The most widely used marker for this purpose is a defective X-linked gene for the enzyme-hypoxanthine guanine phosphoribosyl transferase (HGPRT) (Fig. 6.3). This enzyme functions in the salvage mechanism of nucleotide synthesis whereby the nitrogen bases, hypo-

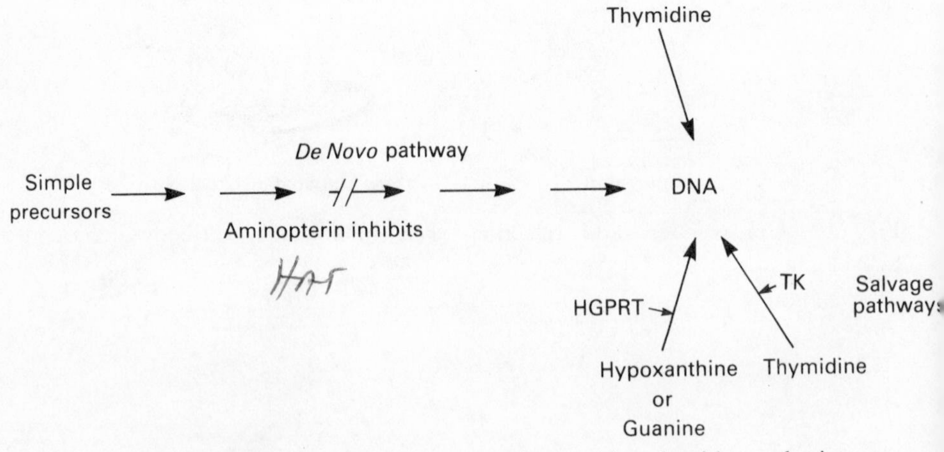

Fig. 6.3 De Novo and salvage pathways of nucleotide synthesis.

xanthine, guanine and thymine are converted to their corresponding nucleotides. The alternative *de novo* pathway for nucleotide synthesis can function by using simple precursors in the absence of this enzyme. Non-mutant cells can use either of these two pathways for nucleotide synthesis. Thus the presence of the metabolic inhibitor, aminopterin, which blocks the *de novo* pathway does not prevent growth of cells which are provided with an adequate supply of the nitrogen bases. However, aminopterin is growth inhibitory to cells having a defective gene — $HGPRT^-$. By choosing myelomas with this gene mutation, aminopterin can be used as a basis for selection of hybridomas from the parental myelomas.

Thus the three important characteristics of myeloma cells used for fusion are (a) infinite life-span, (b) non-production of antibodies, and (c) $HGPRT^-$

(4) The lymphocytes isolated from the spleen of the immunized mouse are mixed with the selected myeloma cells in a medium containing 40–50% polyethylene glycol (PEG). A proportion of the cells will fuse within a minute under such conditions. The process involves cell agglutination, membrane fusion and eventually nuclei fusion.

(5) The fusion treatment results in a mixed population of cells containing — lymphocytes, myelomas and hybridomas. Separation and selection can be carried out in a medium containing HAT — hypoxanthine, aminopterin and thymidine (Table 6.1). This medium is growth inhibitory to $HGPRT^-$ cells but not to $HGPRT^+$ cells which have an active salvage pathway for nucleotide synthesis. The cells are seeded into this medium in the wells of a microtitre plate. The unfused lymphocytes have a limited life-span in culture and soon die. The unfused myelomas, being $HGPRT^-$ will fail to grow in the inhibitory conditions of the HAT medium. Therefore only the $HGPRT^+$ hybridomas capable of continuous growth will survive.

(6) The next stage involves selection of the clones capable of secreting the desired antibody. One method of selection involves binding a fluorescent label to the antigen which is allowed to attach to the hybridomas pro-

Table 6.1 Use of HAT medium

	Cells	
	$HGPRT^+$	*$HGPRT^-$*
De Novo pathway	Inhibited by aminopterin	Inhibited by aminopterin
Salvage pathway	Functional — H & T present in medium	Non-functional — disabled enzyme
Growth	Positive — DNA is synthesized	Negative — no DNA synthesized

ducing the appropriate antibody. Subsequent cell selection requires a fluorescence activated cell sorter which is an instrument capable of recognising the fluorescent stained cells and directing them through a particular channel. This instrument is however expensive and unavailable to many laboratories. An alternative and more usual procedure is to assay the media of isolated clones for the required antibody. Initially, several antibody-producing hybridomas may be isolated but not all are genetically stable owing to the tetraploid chromosomal complement which results in random chromosome loss during the early stages of growth

(7) Re-cloning of the selected hybridoma clones is essential to ensure that the two desired genetic characteristics — continuous growth and antibody production — are stably maintained

(8) These stable hybridomas can be grown in larger quantities by either incorporation into the peritoneal cavity of mice or by cell culture

Large-Scale Antibody Production

In Vivo Incubation

One method of obtaining quantities of a monoclonal antibody from an isolated hybridoma is to inject cells into the peritoneal cavity of the mouse. The animal's abdomen is distended as the cells grow as a tumour in the surrounding ascites fluid which can increase in volume to about 30 ml. Hybridoma cell growth and antibody production is enhanced by endogenous growth factors. Immunosuppressed mice are chosen to avoid immune destruction of the injected hybridomas and contamination of the desired monoclonal antibody with other unwanted immunoglobulins.

Considerations of the advantages and disadvantages of this in vivo method are given in Table 6.2. Although the method has had considerable application on the laboratory scale, it has major disadvantages for the large-scale production of

Table 6.2 Comparison of monoclonal antibody production by *in vivo* and *in vitro* methods

	In vivo	*In vitro*
Amount of Mab produced/ml	1–10 mg/ml	10–200 μg/ml
Volume limitations	5–30 ml/mouse	size of fermenter
Method of operation	multiple units	one unit
Possibilities of manipulating conditions	very limited	wide range
Presence of non-specific mouse Ig	yes	no
Possibility of reducing protein content in crude preparation	no	yes

(From Reuveny, *et al.*, 1985)

monoclonal antibodies. It is a multiple process and requires manipulation of large numbers of mice. This is labour-intensive and could be considered unethical given the availability of alternative methods.

The antibody yield can be as high as 10 mg/ml from the 30 ml of ascites fluid obtained from each mouse. However, this translates to a requirement for over 3000 mice in order to obtain one kilogramme of antibody — clearly an unfeasible operation considering the number of manipulations required. Furthermore, the required antibody is secreted into a pool of non-specific immunoglobulins which must be separated out during the purification process. Therefore, it must be concluded that this *in vivo* cultivation method is unsuitable for large-scale monoclonal antibody production and offers little hope for technological development.

In Vitro *Culture*

Suspension Culture. Hybridoma cells are non-anchorage dependent and can be grown in suspension culture to cell densities up to 2×10^6 cells per ml with an average antibody yield of 100 mg l^{-1}. There are a number of advantages of this method for antibody production as outlined in Table 6.2. The scale of operation can be governed by the size of the fermenter used for the growth of the hybridomas and there are a number of possibilities of improving the efficiency of the process.

Many hybridomas can be adapted to grow in a serum-free medium which contains defined hormones and growth factors. The purification of secreted monoclonal antibodies from such media is much simpler because of the low protein content and absence of non-specific immunoglobulins. At present rates of productivity this method offers the possibility of obtaining a kilogramme of antibody from a single culture of 10 000 l (or 10 culture runs of a 1000 l fermenter).

At Celltech in the UK, airlift fermenters have been found suitable for large scale culture. In this system, a stream of air is pumped into the bottom of a long column

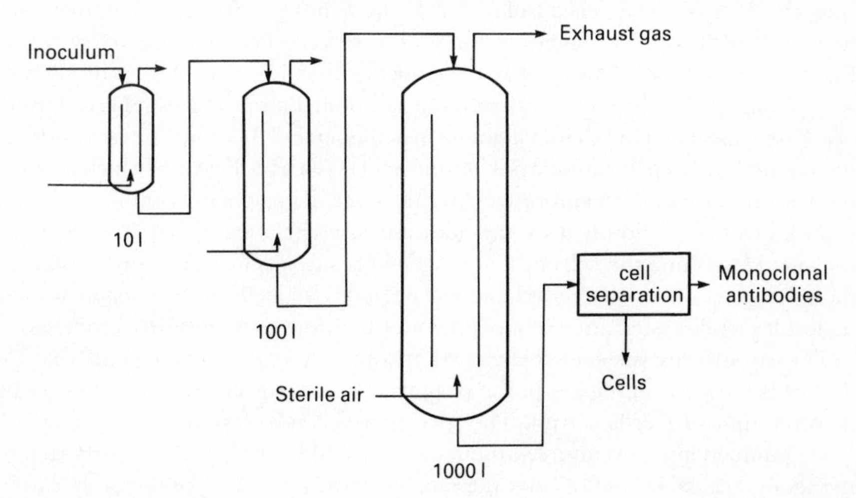

Fig. 6.4 Sequential series of scale-up of hybridoma cultures in airlift fermenters. (From Birch, *et al.*, 1985)

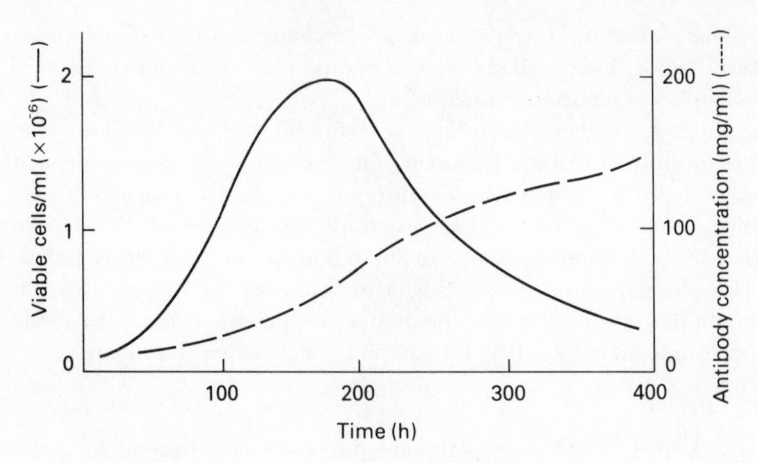

Fig. 6.5 Typical pattern of antibody release from hybridoma cell growth *in vitro*. (From Birch *et al.*, 1985).

in which the cells are held in liquid suspension. The air flow ensures adequate agitation and oxygen supply for growth. Fermenters of this type (up to 1000 l) have been used for monoclonal antibody production. The production procedure is as follows: an inoculum of 1×10^5 cells per ml is allowed to reach 1×10^6 cells per ml in a series of fermenters from 1 to 1000 l. The cell yield from one fermenter is used as the inoculum for the fermenter of the next size, up to 1000 l (Fig. 6.4). Antibody release by the hybridoma cells is maximum at a stationary or declining phase which is maintained as long as possible in the final fermenter in order to maximize the total antibody yield (Fig. 6.5).

Microencapsulation. Another industrial producer of monoclonal antibodies, Damon Biotech, U.S.A., uses encapsulated cells in a process which is commercially named, 'Encapcel'. The stages in the process of cell encapsulation are shown in Fig. 6.6. A high cell density of hybridomas is mixed into a solution of alginate which is allowed to drip into a calcium salt solution. This produces spherical beads of cells encapsulated in calcium alginate of a diameter of 3–5 mm. These beads are then coated with a polycationic polyamino acid (such as polylysine) which acts as a solid semi-permeable membrane. At this stage the alginate contained in the capsules can be re-liquified by the addition of a chelating agent. This leaves a capsule, containing the cells in a homogeneous suspension. The permeability of the polylysine coat can be varied, but will normally allow the free diffusion of small molecules while restricting the movement of both the cells and their products.

The capsules are suspended in growth medium in a simple stirred tank reactor. The cells grow as aggregates in the periphery of the capsule in which they attain densities up to 10^8 cells per ml. The stirring rate can be fast and air or oxygen is freely pumped into the culture without damage to the cells, which are protected by the semi-permeable coat. Under these conditions, up to 20 g of antibody can be produced in a 40 l fermenter in about 20 days — thus requiring a 2000 l culture for a kilogramme of antibody.

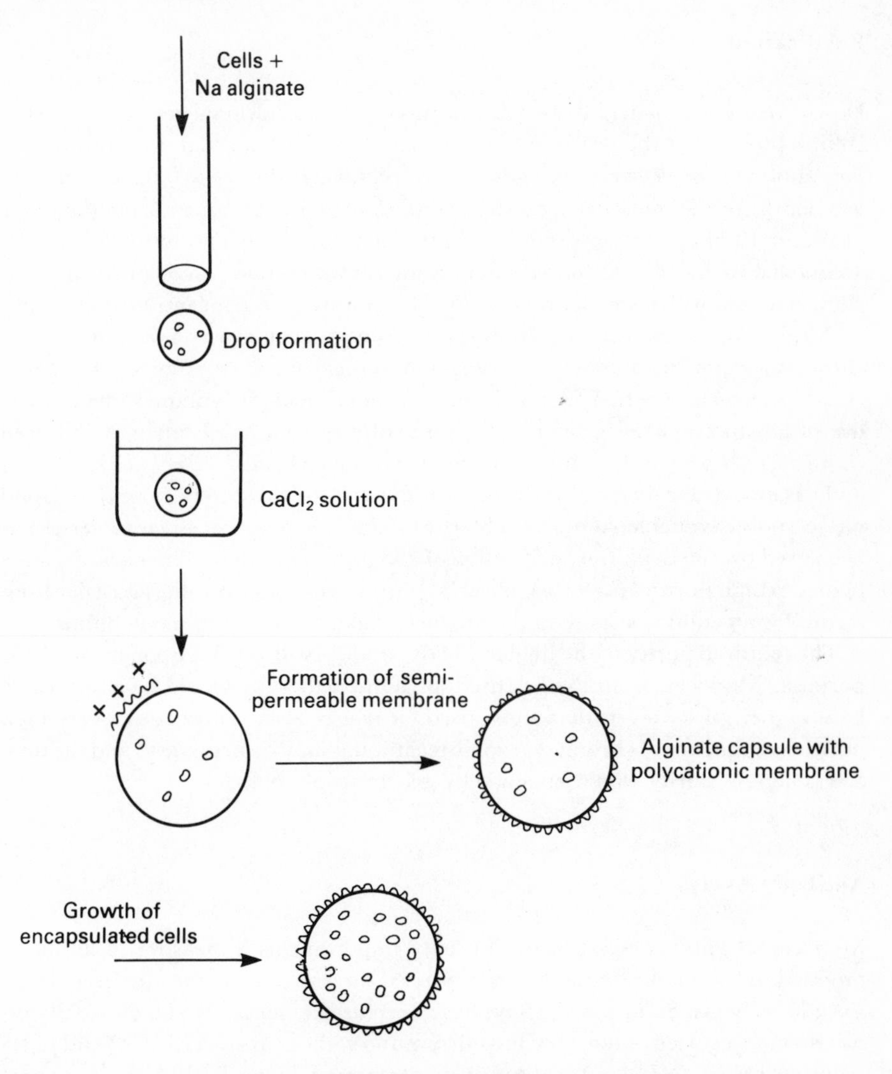

Fig. 6.6 The micro encapsulation process.

The antibody can be isolated by washing the capsules from the growth medium. Gentle homogenization breaks open the capsules without rupturing the cells which may be allowed to sediment out. The resulting supernatant has a relatively high antibody content (up to 75%) which may be purified further by a simple ion exchange column. Problems over the use of serum in the growth medium do not arise because it is physically separated from the secreted antibody. Further purification is therefore relatively easy because of the low content of contaminating proteins in the capsules.

Purification

Downstream processing involves purification of the monoclonal antibody from the culture medium of the processes described above. The expected concentration of the antibody in suspension cultures is relatively dilute — 0.01–0.5 gl^{-1} — whereas from the microencapsulation system, higher concentrations in the order of 0.1 to 10 gl^{-1} may be expected. Initial enrichment of the antibody may be accomplished by ultrafiltration where a membrane retains large molecules but allows the removal of smaller molecules. This process may concentrate up to $\times$ 50.

Purification of the antibody from this concentrate can proceed by a number of chromatographic techniques — adsorption, ion exchange or affinity. One particularly successful method for purification from large media volumes involves the use of an immobilized protein A immunoaffinity column. Protein A is a well characterized protein which is extracted from the cell wall of *Staphylococcus aureus* and has an extraordinary affinity for immunoglobulins. It can be bound to a solid phase and is available commercially as Protein A-Sepharose. Such material has been used routinely for the purification of kilogram quantities of monoclonal antibodies to high purities (>95%). Clearly, purification to such values is easier from serum-free medium because of the absence of extraneous immunoglobulins.

The required purity of the final antibody product will depend upon its intended purpose. Monoclonal antibodies used for human therapy would be expected to have higher purity levels than those used for *in vitro* tests. To isolate to very high purities a sequence of chromatographic techniques may be necessary, and the final assessment of purity would be made by gel electrophoresis.

Antibody Assays

Monoclonal antibody assays are not only important as a measure of antibody production from hybridomas but can be used on a wide scale for the detection of specific antigens. Solid-phase assays have been used extensively and these fall into two main categories — enzyme linked immuno sorbent assays (ELISA) and radio immuno assays (RIA). Quantitative measurement in an ELISA is dependent upon the enzymatic conversion of a substrate to a product which can be detected colorimetrically. RIA involves the measurement of radioactivity — usually ^{125}I which can be bound easily to an antibody.

A schematic representation of a solid-phase binding assay for the detection of a specific antibody is shown in Fig. 6.7. The stages involved in a typical assay are as follows:

(1) A solution of the appropriate antigen is applied to a solid support. This is usually a microtitre plate made of a plastic capable of adsorbing antigens. The adsorption is by non-specific hydrophobic interaction which allows the attachment of most large charged antigens. Difficulties with the binding of small molecular antigens are solved by forming a conjugate

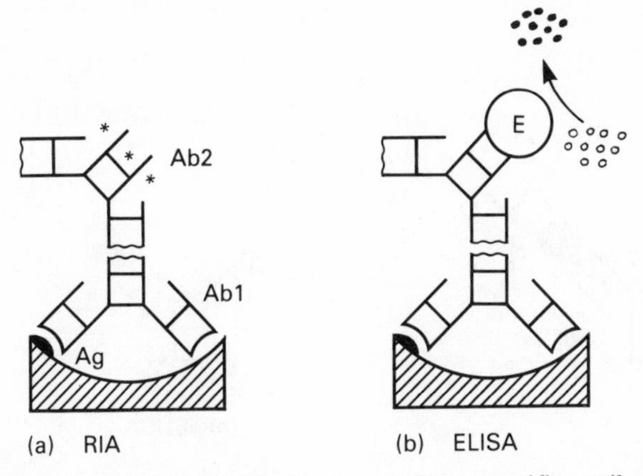

Fig. 6.7 Solid phase binding assay for a specific antibody.
(Re-drawn from Campbell, 1984).

with a larger carrier protein such as bovine serum albumin (BSA) which does not interfere with the assay

(2) The remaining attachment sites on the solid support are blocked by the addition of a non-interfering protein — again BSA is often used. This is to ensure that there is no possibility of antibody attachment to the solid support at later stages

(3) A solution of the monoclonal antibody under test is added. This will bind to the antigen which is held on the solid support

(4) The degree of attachment of this monoclonal antibody is assessed by a second antibody. This second antibody has a specificity against the total Ig of the species from which the hybridoma was produced. For example, rabbit anti-mouse Ig can be used if the first antibody was derived from a mouse. The second antibody has an enzyme (for ELISA) or a radiolabel (for RIA) covalently attached. Conjugated anti-Ig antibodies of various types are readily available from commercial sources

(5) In RIA, measurements of radioactivity can be made by autoradiography or by scintillation counting of the solid support

(6) In ELISA, the appropriate substrate is added before colorimetric measurement of its product. Three enzymes have been widely used for ELISA — β-galactosidase, alkaline phosphatase and horse radish peroxidase. These have been chosen because of colour changes associated with their enzymatic activity. For example, a colourless solution of p-nitrophenyl phosphate can be converted to the yellow p-nitrophenol by alkaline phosphatase. The colour intensity can be measured on a micro-titre plate by the use of a multiscanning spectrophotometer

These procedures for antibody assay described above can be conveniently performed in a multiwell microtitre plate in which solutions can be easily added and rinsed out. In such an assay it is usual to include various dilutions of the antigen and the antibody.

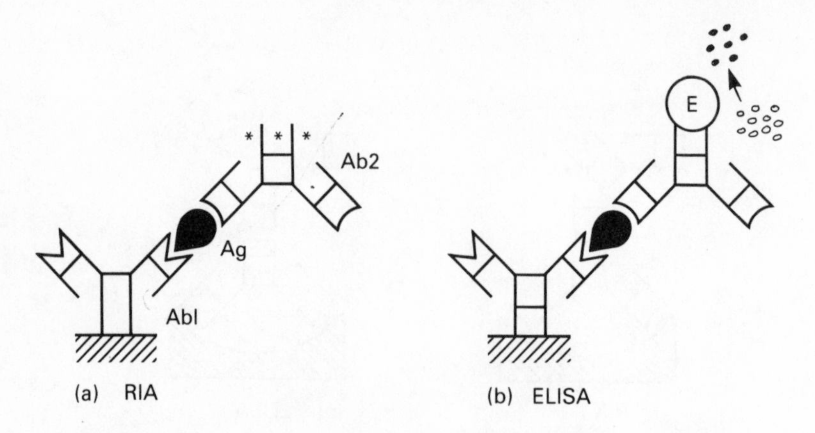

Fig. 6.8 Solid phase binding assay for a specific antigen.
(Re-drawn from Campbell, 1984).

The stages of a binding assay for a specific antigen would be performed in a different order from that described earlier. Figure 6.8 shows the basis of a typical sandwich assay in which the first antibody is held on the solid support and the antigen is bound by both the first and second antibody. The antibodies in this case are directed against two different epitopes of the same antigen. This ensures extremely high specificity and can be used, for example, to distinguish between interferon sub-types (see Chapter 5).

Uses of Monoclonal Antibodies

The high selectivity of binding of monoclonal antibodies to specific antigens has enabled a range of applications. The monoclonal antibodies can be used to identify or isolate minute quantities of specific antigens in the presence of a bulk of components which may or may not have chemical similarities to the antigen. This has been particularly useful in medical diagnosis for which monoclonal antibody assay kits are available for identifying viruses, bacteria or parasites associated with a range of specific diseases. Also, increases in enzyme levels associated with such conditions as coronary disease or brain damage can be detected. Immunodiagnostic tests of this type can be performed within 15–20 min — such speed is of considerable benefit for early treatment. Not only can the presence of a virus be detected in body fluids but genetic variants of the same basic strain can be distinguished. This allows the epidemiological study of a disease, knowledge of the location and origin of which can prevent its spread.

The diagnostic applications of monoclonal antibodies are rapidly expanding. Considerable interest is being shown in the use of specific radiolabelled antibodies as diagnostic probes for tumours, which could lead to simple means of mass screening of certain types of cancers. Some of these antibodies may be required in extremely large quantities, such as the ABO blood typing reagents, which have production levels predicted in kilogram quantities per year.

The potential therapeutic uses of monoclonal antibodies are considerable. Mouse monoclonal antibodies have been used for immunosuppression in human organ transplant operations and have been successful in avoiding the problems of tissue rejection. However, in many cases of therapeutic use, mouse monoclonal antibodies are unsuitable because antibodies against mouse Ig are produced in the treated patients. Techniques for the routine production of human monoclonal antibodies are in the developmental stages and the general availability of these will allow a much wider use in human therapy. One particularly significant development is targeted drug therapy with 'magic bullets'. For example, antibodies to surface antigens of cancer cells may be conjugated to cytotoxins, e.g. ricin or diptheria toxin. The injection of such a conjugate will allow targeting of the cancerous cells, around which will be produced a high concentration of the cytotoxin leading to the destruction of these cells. Although this work is presently in the experimental stage, the routine use of such techniques may be anticipated in the near future.

The development of antibodies which combine with the antigen binding site of other antibodies are called anti-idiotype antibodies. They have complementary specificity to an antibody and similarities to the original antigen. These can act as pseudo-antigens and will elicit a response from the immune system. The value of such anti-idiotype antibodies as passive vaccines is considerable, particularly for problematic diseases such as rabies (see Chapter 4).

The use of monoclonal antibodies for chromatographic procedures can be conducted on an industrial scale for the purification of such compounds as interferon (see Chapter 5). Because of the high specificity of binding, small chromatographic columns can be used repeatedly to extract low concentrations of an antigen from large quantities of raw material.

Human Monoclonal Antibodies

The techniques that have been described so far in this chapter have been developed in experimental animals — particularly mice. However, the major problem with the use of murine cell-derived monoclonal antibodies for human disease treatment is that antibodies can be raised against them in the recipients which can both limit their effectiveness against the target antigen and may induce hypersensitivity due to the presence of immune complexes. Therefore, the ability to produce large quantities of human monoclonal antibodies would have advantages particularly for therapeutic use.

There are two major difficulties associated with the production of cells capable of secreting human monoclonal antibodies:

(1) The source of antibody-secreting lymphocytes. In developing mouse hybridomas, the macerated spleen of immunized animals is used as the

source of the mixed lymphocyte population from which cell selection can be made. For human lymphocytes, the source is largely limited to blood samples taken from patients who have acquired an infection and developed an immunity against the targeted disease. Often low specific antibody production is found in the lymphocytes isolated from such samples

(2) Immortalization and chromosome stability. The frequency of cell fusion of human lymphocytes with human lymphoid tumour cells is low compared with equivalent fusions in mouse cells. Also, the resulting hybridomas are less stable and tend to lose their chromosomes. Mouse–human hybrids for human monoclonal antibody production can be made, but these often preferentially lose human chromosomes after a limited number of generations in culture

There are several ways to alleviate these problems. The procedures for immortalization of the cells have been extensively investigated in order to increase the frequency of production of human antibody-secreting hybridomas. The list of continuous growth human cells suitable as fusion partners is increasing and the lymphoblastoid or myeloma cells isolated from patients have a varying potential for successful fusions. Mutants can be induced to improve fusion ability or to incorporate genetic markers.

Cell transformation has been considered as an alternative to cell fusion. Infection with Epstein Barr virus can transform cells to continuous cell lines capable of infinite growth capacity. Also, transfection with tumour-derived DNA has been used to good effect. However, any products derived from such transformed or transfected cells would have to be fully purified from viral or DNA contamination before human use.

The possibilities of stimulating lymphocytes to produce antibodies to selected antigens *in vitro* is being investigated. If conditions could be optimized for such lymphocyte stimulation in culture then the whole process of cell selection of specific antibody producers would be considerably improved. Furthermore, such a method, if perfected, would abrogate the need to use animals for immunisation which is costly and time consuming and sometimes unpredictable.

Although the technology is not yet fully developed for the large-scale production of human monoclonal antibodies there is no doubt that further research and development of the techniques mentioned here will ensure that these will be the next generation of commercially produced monoclonal antibodies.

Conclusion

Hybridoma technology has developed considerably from the first experimental report of monoclonal antibody production in 1975. Many biotechnology companies have now dedicated their activities to their production on a routine basis. In particular, tests for blood typing and diagnosis of a range of diseases can now be performed routinely by the use of commercially produced monoclonal antibodies — over 70 have now been approved for use in diagnostic immunoassays. Because of the extent of such applications, kilogram quantities of certain monoclonal antibodies are being regularly produced. There is no reason to believe that this trend for diagnostic uses will not continue.

The use of monoclonal antibodies for therapy is limited by the availability of those of human origin. This is almost certainly the next application which will see major developments resulting from experimental techniques for the production of human–human hybridomas. The use of anti-idiotype antibodies as vaccines is in an early stage of development, but this too is likely to result in significant future applications.

Future benefits could be gained by improved techniques for immunization and selection of antibody-secreting populations of lymphocytes. This may come about by the development of *in vitro* lymphocyte stimulation techniques which could dramatically increase the variety of monoclonal antibodies generally available.

Summary

Hybridomas are fused cells which can be constructed for continuous growth and monoclonal antibody production. The techniques for such hybridoma production were reported by Kohler and Milstein in 1975 and involve the fusion of an antibody-secreting B-lymphocyte with an immortal myeloma. The careful selection of a hybridoma with the desired properties can lead to the production of a clone of cells capable of indefinite production of a desired monoclonal antibody. Commercial methods for large-scale growth have included suspension cultures which have been particularly successful in airlift fermenters. Some advantages have also been found for the immobilization of the hybridomas in gel capsules which can accumulate the cell-secreted antibodies.

The value of a monoclonal antibody lies in the high specificity shown in binding compounds of choice. This can be used for the identification of the compounds or cells, e.g. in blood typing or disease diagnosis. The assays involved are simple and quick to perform.

Although most commercial monoclonal antibodies to date have been produced from murine hybridomas, future developments will undoubtedly result in routine production of human hybridomas capable of synthesizing human monoclonal antibodies. These developments will in turn lead to the increased use of monoclonal antibodies for human disease therapy including cancer treatment.

General Reading

Cambell, A.M. (1984). *Monoclonal Antibody Technology*. Barking, Elsevier Applied Science.
Kozbor, D. and Roder, J.C. (1983). 'The production of monoclonal antibodies from

human lymphocytes,' *Immunology Today* **4**, pp. 72–79.
Osterhaus, A. and Uytdehaag, F. (1985). 'Lymphocyte hybridomas: production and use of monoclonal antibodies,' in *Animal Cell Biotechnology*. vol. 2 (Eds Spier R.E. and Griffiths B.) London, Academic Press pp. 49–69.
Sikora, K. and Smedley, H.M. (1984). *Monoclonal Antibodies*. Oxford, Blackwell Scientific.

Specific Reading

Amzel, L.M. and Poljak, R.J. (1979). 'Three-dimensional structure of immunoglobulins', *Ann. Rev. Biochem.* **48**, pp. 961–967.
Birch, J.R. *et al.* (1985). 'Bulk production of monoclonal antibodies in fermenters', *Trends in Biotech.* **3**, pp. 162–166.
Capra, J.D. and Edmundson, A.B. (1977). 'The antibody combining site,' *Sci Amer.* **236**, pp. 50–59.
Duff, R.G. (1985). 'Microencapsulation technology: a novel method for monoclonal antibody production', *Trends in Biotech.* **3**, pp. 167–170.
Kennedy, R.C. *et al.* (1986). 'Anti-idiotypes and immunity', *Sci. Amer.* **255**, pp. 40–48.
Kohler, G. and Milstein, C. (1977). 'Continuous cultures of fused cells secreting antibody of predefined specificity', *Nature* **256**, 495–497.
Lord, J.M. *et al.* (1985). 'Immunotoxins', *Trends in Biotech.* **3**, pp. 175–179.
Milstein, C. (1980). 'Monoclonal antibodies', *Sci Amer.* **243**, pp. 66–74.
Nowinski, R.C. *et al.* (1983). 'Monoclonal antibodies for diagnosis of infectious diseases in humans', *Science* **219**, pp. 637–644.
Posillico, E.G. (1986). 'Microencapsulation technology for large-scale antibody production', *Biotechnology* **4**, pp. 114–117.
Randerson, D. (1984). 'Hybridoma technology and the process engineer', *Chemical Engineer*, No. 409.
Scott, M.G. (1985). 'Monoclonal antibodies — approaching adolescence in diagnostic immunoassays', *Trends in Biotech.* **3**, pp. 170–175.
Secher, D.S. and Burke, D.C. (1980). 'A monoclonal antibody for large-scale purification of human leucocyte interferon', *Nature* **285**, pp. 446–450.
Yelton, D.E. and Scharff, M.D. (1981). 'Monoclonal antibodies: a powerful new tool in biology and medicine', *Ann. Rev. Biochem.* **50**, pp. 657–680.

Chapter 7

Insulin

Introduction

The regulation of mammalian blood sugar levels is controlled by two hormones which are synthesized in clusters of cells scattered throughout the pancreas. These clusters are the islets of Langerhans, which have cells of two types — the A and B cells, which synthesize glucagon and insulin respectively. The synthesis of each hormone can be induced by the glucose level of the blood. Concentrations of blood glucose >90 mg/100ml of blood stimulate insulin production whereas levels <80 mg/100ml stimulate glucagon production. Each hormone is secreted into the blood stream where the physiological effects tend to restore the glucose level to a norm of 80–90 mg/100ml.

Glucagon acts primarily on liver cells, and through a cell surface receptor causes glycogen degradation (glycogenolysis). Insulin also acts through cell surface receptors mainly present in muscle and adipocyte cells. Its major effect is to stimulate the activity of glucose permease enzymes which serve to increase the permeability of these cells to glucose. However, there are a number of other metabolic activities of insulin, including (a) the promotion of cellular uptake of amino acids, (b) an increase in the rate of protein and lipid synthesis, and (c) inhibition of lipolysis and gluconeogenesis. All these changes can be brought about by relatively small amounts of insulin acting through a cAMP-dependent protein kinase, although the detailed mechanism of this is not clear. The concentration of insulin normally circulating in the blood stream varies from 0.2 ng/ml to 2ng/ml.

Diabetes Mellitus

The disease diabetes mellitus is named from the observed symptom of excessive urination of sweetened (honey-like) urine. Associated with this is a high blood sugar level which results from a decreased ability of cells to absorb glucose. The condition affects 5% of the population and is often due to a deficiency of insulin.

In 1922 Best and Banting established the basis of our undertanding and treatment of diabetes through a set of experiments in which diabetes was induced by pancreatectomy (removal of the pancreas) in experimental animals. They found that this condition could be alleviated by subsequent injections of an extract taken from canine pancreases in which the secretory ducts had been ligated for up to 10 weeks earlier. The extract, which was initially made into cold Ringers solution and filtered, could consistently reduce the blood glucose level of depancreatised dogs.

Results resembling pancreatectomy can also be observed by administration of alloxan, which is a chemical capable of selectively destroying the B cells of the pancreas. Subsequent injection of aqueous or ethanolic pancreatic extracts into these animals can also alleviate the diabetic condition. Best and Banting established that the active ingredient of this pancreatic extract was derived from the cell clusters of the islets of Langerhans and so derived the name, insulin. Subsequent purification and analysis of the active pancreatic extract identified a small peptide which accounts for all the anti-diabetic effects. Later it was shown that such extracts could be used in the treatment of diabetic patients.

Human diabetes may result from the inability of the sufferer to synthesize normal insulin. This can be an inherited condition or may arise through viral, chemical damage to the secreting cells in the pancreas or through a genetic mutation resulting in the production of a defective insulin molecule. These cases can be treated by regular injections of insulin — substitution therapy.

Structure of Insulin

Insulin was selected by Sanger in the 1950s as the first polypeptide for complete analysis by his newly developed technique of protein sequencing. The choice of insulin was based on its relatively small size. Figure 7.1 shows that the two peptide chains — A (21 amino acid residues) and B (30 amino acid residues) — are held together by two disulphide bonds. Separation of the two chains can be affected by oxidation of these disulphide bridges. However, once separated the re-establishment of the biological activity of insulin is difficult because it requires that the disulphide bonds are fully restored to their original conformation. This can be problematic because of the various alternative configurations that can be assumed.

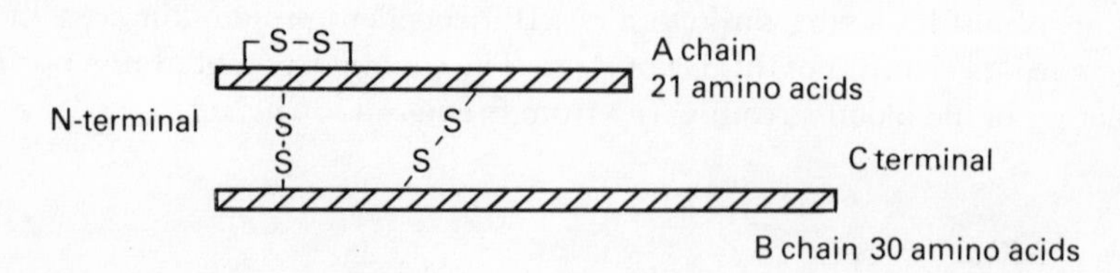

Fig. 7.1 Structure of insulin.

Normal Physiological Synthesis

Under normal physiological conditions insulin is synthesized in B cells of the human pancreas. The first protein to be synthesized from the insulin gene is a single polypeptide precursor of 110 amino acid residues called pre-proinsulin. This consists of sequences corresponding to the A and B chains of the mature insulin with an intervening sequence designated C (35 amino acids) and an extra N-terminal peptide (23 amino acids) (Fig. 7.2). Soon after synthesis, this 23 amino

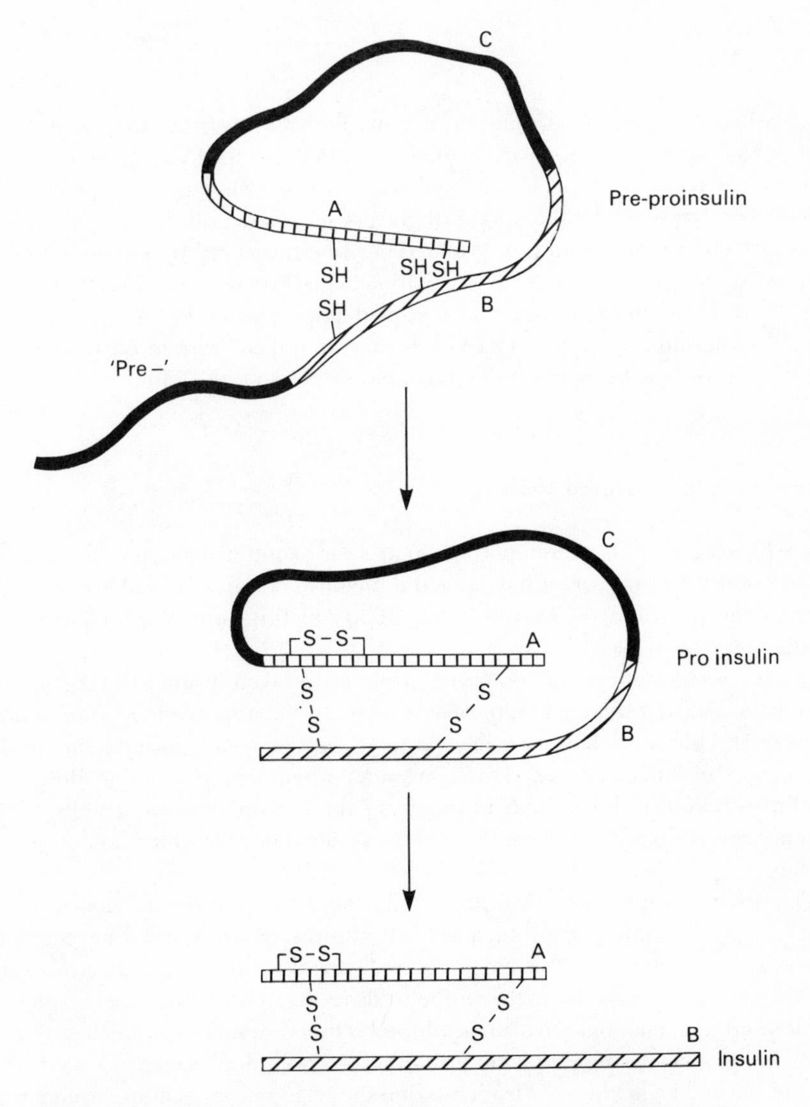

Fig. 7.2 Physiological synthesis of insulin in pancreatic cells.

acid 'pre-' peptide is removed by proteolytic cleavage and the three disulphide bridges are formed.

It is important to notice that the C chain ensures the folding and orientation of the A and B chains so that the disulphide bridges can be formed correctly. The pro-insulin is transported to the Golgi apparatus where further proteolytic cleavage removes the C chain. The resulting mature insulin is then packaged in vesicles for secretion. There are no additional post-translational modifications such as glycosylation and the protein is released as a 51 amino acid residue dipeptide.

In Vitro Assay

The earliest methods of assessing the activity of insulin was through its ability to reduce blood glucose levels in experimental animals — the hypoglycaemic effect. The use of isolated tissue has proved more acceptable. For example, insulin activity can be shown by the uptake of glucose in isolated rat diaphragm muscle. Alternatively, the promotion of glycolysis can be measured in adipose tissue by monitoring the conversion of radioactivity labelled glucose to carbon dioxide.

More recently, immunoassays have proved quicker and more effective, and the use of radioimmune assays (RIA) or enzyme-linked immunosorbent assays (ELISA) have now become routine for rapid detection (see Chapter 6).

Secretion from Cultured Cells

There have been various attempts on a small scale to obtain insulin from the *in vitro* culture of cells. However, it has proved difficult to obtain reasonable concentrations of the peptide to consider the possibility of this approach for large-scale production.

Insulin secretion can be observed from cells taken from primary animal pancreatic cultures and the rate of release may be improved by stimulation. Glucose and glucagon have been shown to act synergistically in inducing insulin production in such cultures. However, after about ten population doublings insulin secretion declines. This results from the inevitable heterogeneity of the primary cell population and consequent overgrowth of cells which do not secrete insulin.

The ability to produce a cell line exhibiting a stable differentiated function depends upon cloning from such mixed cultures. Such a cell line could be hybridized to a transformed cell to produce a hybrid capable of infinite growth and insulin secretion — similar to the antibody-secreting hybridomas (see Chapter 6).

Alternatively, the isolation of an insulin-secreting tumour cell would be valuable in obtaining a means of permanent insulin production. Such a transplantable islet cell tumour has been obtained from rats and shown to release insulin through continuous growth in culture. However, the rate of insulin release (5–25 ng/ml of culture/day) is not sufficient to merit culture of such cells for large-scale insulin production.

Therapeutic Use

The structure of insulin differs only slightly between species. Porcine insulin (derived from pig) differs from human insulin by only one amino acid. The C terminal amino acid of the B chain is an alanine in porcine insulin and threonine in human insulin.

For many years, supplies of insulin for diabetic treatment have involved large-scale extraction from pig pancreas. The purified insulin from this source has been found suitable for therapeutic use in the majority of cases. However, two problems have arisen to suggest the need for an alternative source:

(1) Porcine insulin causes an immunogenic response in a small percentage of diabetic sufferers because of the alternate amino acid in the structure
(2) The supply of animal pancreases has been vulnerable to fluctuations in the meat trade which has often failed to provide sufficient material for extraction. The situation is exacerbated by the increased demand for insulin which has arisen because of the increased life expectancy of diabetics

These problems argue for a constant supply of human insulin.

Production of Semi-Synthetic Human Insulin

Although the chemical synthesis of insulin is possible because of its relatively small size, such an operation would be too time- consuming and expensive for routine large-scale production. However, chemical conversion of extracted porcine insulin to human insulin is feasible. This of course solves the problem of immunoreaction against treatment but does still require a good supply of pig pancreas for initial isolation. The Danish company, Novo, is undertaking the manufacture of human insulin by such a chemical conversion from extracted pig insulin.

The first method developed for conversion of pig to human insulin was reported in 1972 by Ruttenberg, who suggested a protocol based on the removal of a fragment of the B chain of porcine insulin and replacement by a synthesized short peptide according to the known sequence in human insulin. In order to allow such a change, a series of blocking agents had to be used so as to ensure the correct orientation of the incoming peptide. The stages involved in Ruttenberg's chemical conversion are shown in Fig. 7.3 and are as follows:

(1) Porcine insulin is treated with diazomethane to methylate all the free carboxyl groups. This includes 4 γ-carboxyl groups of glutamic acid and 2 C terminal peptide chain carboxyls. For simplicity only methylated C terminal groups are shown in Fig. 7.3

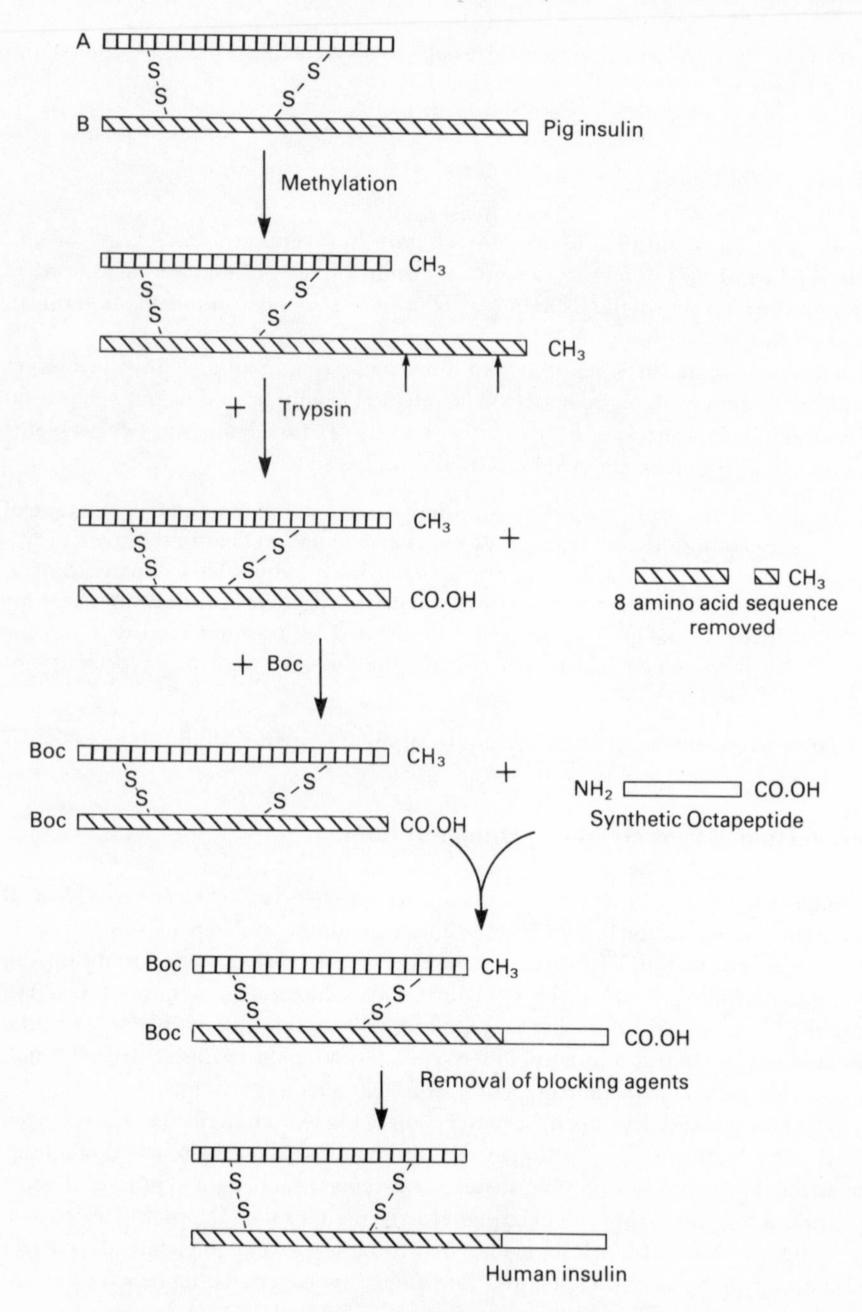

Fig. 7.3 Synthesis of semi-synthetic human insulin.

(2) The methylated insulin is treated with trypsin which cleaves peptide bonds if lysine or arginine occur at the C end. This allows two breaks at the C end of the B chain of insulin and eliminates 8 amino acids leaving an arginine residue at position 22 with an exposed carboxyl group. This modified molecule is named desoctapeptide (B23–B30) insulin
(3) This molecule is treated with tertiary butyl oxycarboxyl (BOC) azide which blocks all free amino groups with BOC. Figure 7.3 shows the blocked N terminal amino groups.
(4) A synthetic octapeptide corresponding to residues 27 to 30 of human insulin is chemically synthesized from its amino acids. This octapeptide is mixed with the previously treated and blocked insulin residue. A coupling reaction in this mixture allows the octapeptide to condense with the only available group, which is the C terminal of the B chain
(5) The blocking agents are removed by hydrolysis and the resulting human insulin purified

During this process the disulphide groups remain intact and therefore there is no problem of re-establishing the relative positions of the two chains.

Alternative methods based on conversions of porcine insulin have involved enzymic reactions which allow transformation of the terminal amino acid residue. In one such process, carboxypeptidase Y (from yeast) under carefully controlled reactions can be used to remove the C terminal alanine by deamidation and then, in a second reaction step, promote the addition of threonine to the separated product. Although such an enzymic conversion sounds facile, the problem lies in the number of alternative products formed. These must be carefully separated and purified by chromatography — usually high pressure liquid chromatography (HPLC). The yield of the final product is at best 30%. However, the process does preclude the need for chemical blocking and de-blocking reactions.

Production of Human Insulin from Genetically Engineered Bacterial Cells

Because of its small size and therapeutic need, human insulin was one of the first molecules to be considered for production by genetically engineered bacteria. Most of the work in construction of suitable insulin-producing clones was performed by Goeddel *et al.* at Genentech in the U.S.A. and reported in 1979.

Their first approach was to consider synthesis of the A and B chains separately. Rather than isolate authentic human insulin genes or transcripts they used a method for the construction of entirely synthetic genes. The 21 and 30 amino acid residues of the A and B chains require nucleotide sequences of 63 and 90 which were short enough to be considered for chemical synthesis. From the known amino acid sequence of human insulin a suitable nucleotide sequence was worked out from the 'preferred' codons in *E. coli*. Such a sequence was not identical with the authentic human gene — rather it was based on the triplet codons which are known to be used preferentially in *E. coli*.

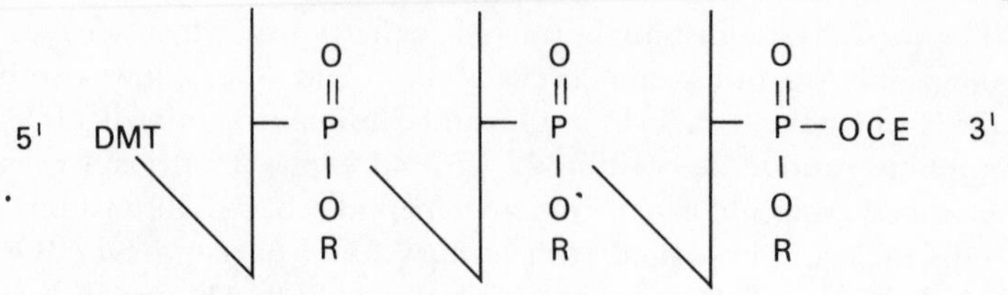

5' DMT — O=P(OR) — O=P(OR) — O=P(OR)—OCE 3'

DMT = Dimethyl oxytrityl

CE = β-cyanoethyl

Fig. 7.4 A blocked trimer used in the synthesis of nucleotide chains.

The experimental approach taken was as follows:

(1) Block-coupling methods were used to construct synthetic oligonucleotides of 10 to 15 units in length. These were synthesized from trimers (trinucleotides) corresponding to pre-determined codons (Fig. 7.4). The reactive 5′ and 3′ ends of each trimer can be blocked by specific reagents. Hydrolysis of these blocked trimers can be initiated in either one of two ways, to selectively remove either of the blocking agents. The removal of one of the blocking agents from the ends of each of the trimers can ensure that the condensation to an incoming trimer occurs in the required orientation

(2) The synthetic nucleotide fragments — deca- to pentadeca-nucleotides — were purified by HPLC. Then in a sequential series of condensation reactions involving selective removal of blocking agents they were joined to form two synthetic genes for the A and B chains of insulin

(3) Each of the two synthetic nucleotide sequences was further modified by the addition of a codon for the amino acid (methionine) at the start of the insulin sequence and a terminator at the end (Fig. 7.5)

(4) By the use of cohesive linkers, the constructed genes were inserted independently into the plasmid, pBR322 which already contained the β-galactosidase gene. This gene has a promoter sequence that can allow protein expression in procaryotic cells. Care was taken so that the synthetic genes were inserted into the normal reading frame of the β-galactosidase gene, downstream of the promoter sequence

(5) Transformation of *E. coli* by the recombinant plasmids resulted in independent clones capable of synthesizing either chain A or chain B of human insulin

(6) Extraction of the synthesized peptides from these clones were attempted from the bacterial cell lysates. The product consists of a fusion protein with part of the β-galactosidase amino acid sequence joined to either chain A or B of insulin. On isolation of the fusion proteins, the insulin peptides can be recovered by treatment with cyanogen bromide. This cleaves peptides at methionine or tryptophan positions and as neither of these

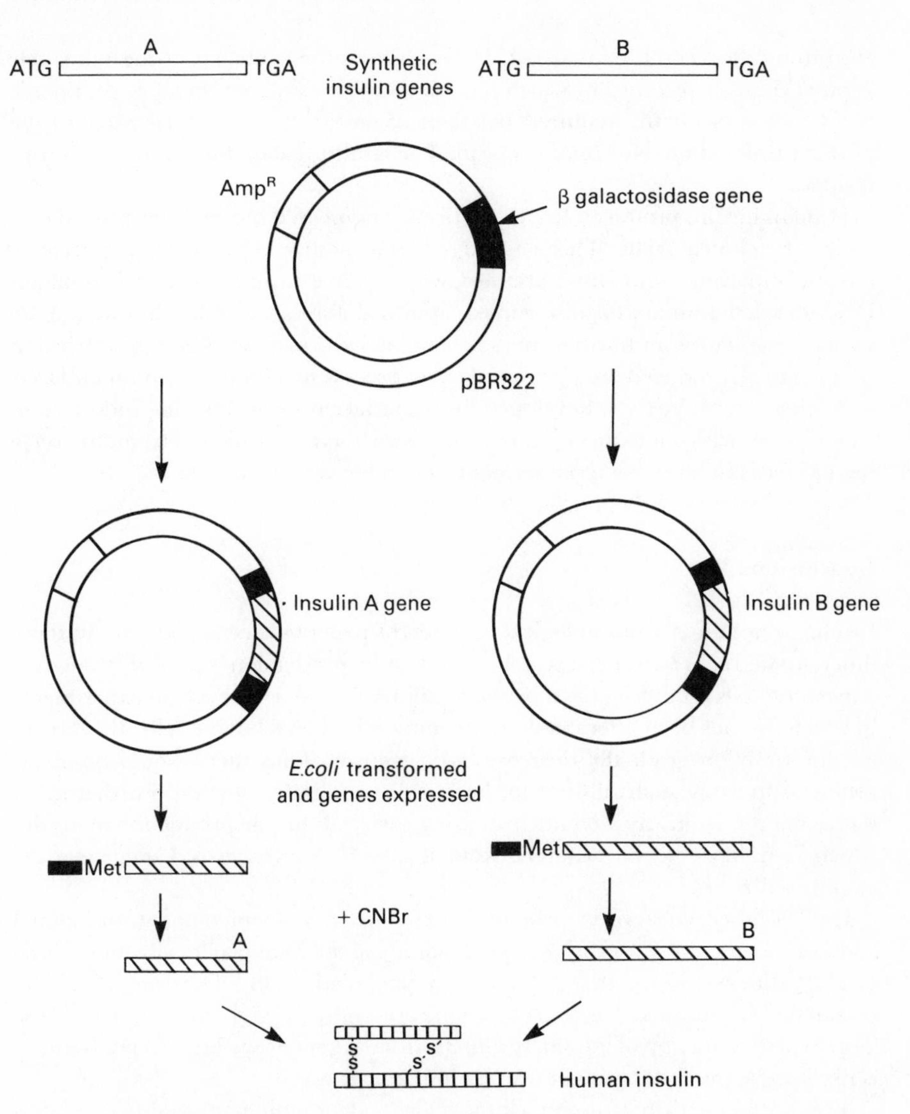

Fig. 7.5 Production of human insulin in bacteria using synthetic genes.

amino acids are present in insulin, the only point of cleavage is between the β-galactosidase and insulin chain

(7) The A and B chains were purified separately and then mixed in a reaction to reconstitute the disulphide bridges. This latter reaction can result in the formation of disulphide bonds with various orientations and the yield of the authentic insulin molecule with full biological activity can be low.

A second approach to synthesize insulin from genetically engineered cells involved the synthesis of an artificial pro-insulin gene which was inserted into *E.*

coli through a recombinant plasmid. This allows the synthesis of the full single peptide chain of pro-insulin which has the correct orientation to allow disulphide bond formation at the required positions. Selective enzymatic cleavage of the purified product enables removal of the C chain to release the authentic human insulin.

Human insulin produced from genetically engineered bacteria has proved efficacious by clinical trials. This was judged by its ability to lower blood glucose to levels comparable with those attained with porcine insulin. Such recombinant DNA-derived human insulin gained approval for large-scale clinical use by various regulatory authorities in 1982, and allowed insulin to be the first health product to be produced on a large scale by a process involving recombinant DNA technology. Eli Lilly (who developed the bacterial clones for insulin production in conjunction with Genentech) now synthesize their so called 'Humulin' on a regular basis in large bacterial fermenters at a capacity of 10 000 l.

Conclusion

Insulin secretion is a physiological function of pancreatic cells and only in these differentiated cells is the gene switched on. Insulin production is possible from such pancreatic cells established as a primary culture *in vitro*. However, in experiments in which this has been attempted, for example with foetal bovine cells, the level of insulin secretion gradually decreases with culture. Thus there would be major problems in using such cultures for large-scale insulin production. Furthermore, there does not seem any advantage in using such cells for the production of insulin which is unlikely to be different from the product synthesized in genetically engineered *E. coli*.

Unlike other eucaryotic proteins, there are no obvious post-translational additions or modifications to insulin which make the final genetically engineered product different from that produced in pancreatic cells. The rate of insulin production in cultures of genetically engineered *E. coli* is high. Consequently, the large-scale production of human insulin from such genetically engineered bacteria is likely to be the predominant process in the near future.

The continued use of the semi-synthetic method for human insulin production is dependent upon the fluctuating cost and availability of the porcine insulin used as starting material. The chemical or enzymatic conversion to human insulin is effective and so such a production method may also be used for a number of years to come.

A promising development may result from the idea of constructing an artificial pancreas capable of hormone secretion. This may be possible by the encapsulation of appropriately differentiated cells in a semi-permeable matrix which can be inplanted into patients. The permeability of such a chamber should allow the movement of nutrients and secreted hormones but prevent the release of cells or intake of large proteins such as antibodies. The advantage of such an artificial pancreas for therapy is that hormone release (insulin or glucagon) may then be responsive to the prevailing blood glucose levels. Alternatively, regular or slow continuous hormone release from such a system would be preferable to daily insulin injections because the blood insulin would then be maintained at a relatively stable concentration. Such systems have been shown to work in experimental animals.

Summary

The successful treatment of some forms of diabetes has been possible for many years by regular injections of insulin. This small peptide hormone is secreted from specific cells in the pancreas in response to certain stimuli, notably blood glucose and glucagon. Such secretion is defective in some forms of diabetes.

Insulin is a relatively small molecule consisting of two peptide chains held together by disulphide bridges. The structure does not show appreciable species differences. Animal insulin has been used for human diabetes treatment — particularly pig insulin — which differs from the human type by one amino acid. However, the unreliable supply of pig pancreases (and some cases of immunoreaction against the extracted insulin) has led to the need for large-scale production of authentic human insulin.

One approach to obtaining human insulin is by chemical conversion from pig insulin. This involves the conversion of a terminal alanine residue to threonine and can be accomplished by the combined use of enzymes and peptide synthesis. During the process, the bonds between the two peptides are not broken and the resulting semi-synthetic product is identical to human insulin.

The synthesis of human insulin from genetically engineered bacteria offers a means of large-scale production. Two approaches have been taken in the construction of suitable genetically engineered strains. In the first approach the genes for the two individual peptides of insulin were synthesized from trimer building blocks. These were inserted separately into plasmids which were used to transform *E. coli*. The two insulin peptides were synthesized separately from selected strains. Controlled chemical conditions will allow the two peptides to link together, thus producing an insulin molecule identical to the authentic product from human pancreatic cells. A second approach to producing human insulin involves synthesis of the full pro-insulin gene sequence. This consists of a long polypeptide chain which normally undergoes proteolytic cleavage in mammalian cells before secretion. The pro-insulin molecule isolated from selected genetically engineered bacteria is treated with proteolytic enzymes to enable production of the authentic human product. In both cases, the human insulin derived from recombinant DNA has proved efficacious in clinical trials.

The methods described allow large-scale production of human insulin by semi-synthesis from pig insulin or synthesis from genetically engineered bacteria. These methods are far more economically feasible than production from isolated mammalian pancreatic cells which can be induced to synthesize insulin. However, future developments may allow the construction of artificial pancreases based on mammalian cells enclosed in semi-permeable chambers. Such a system has many advantages in the therapeutic treatment of diabetics compared with regular insulin injections.

Specific Reading

Chick, W.L. *et al.* (1977). 'Artificial pancreas using living beta cells: effects of glucose homeostasis in diabetic rats', *Science* **197**, pp. 780–781.

Crea, R. *et al.* (1978). 'Chemical synthesis of genes for human insulin', *Proc Nat. Acad. Sci.* **75**, pp. 5765–5769.

Emerick, A.W. *et al.* (1984). 'Expression of a β-lactamase preproinsulin fusion protein in *E. coli*', *Biotechnology* **2**, pp. 165–168.

Goeddel, D.V. *et al.* (1979). 'Expression in *E. coli* of chemically synthesized genes for human insulin', *Proc. Nat. Acad. Sci.* **76**, pp. 106–110.

Johnson, I.S. (1983). 'Human insulin from recombinant DNA technology', *Science* **219**, 632.

Lim, F. and Sun, A.M. (1980). 'Microencapsulated islets as bioartificial endocrine pancreas', *Science* **210**, 908.

Morihara, K. *et al.* (1979). 'Semi-synthesis of human insulin by trypsin-catalysed replacement of Ala-B30 by Thr in porcine insulin', *Nature* **280**, pp. 412–413.

Mosbach, K. *et al.* (1983). 'Formation of proinsulin by immobilized *Bacillus subtilis*', *Nature* **302**, pp. 543–545.

Ruttenberg, M.A. (1972). 'Human insulin. Facile synthesis by modification of porcine insulin', *Science* **177**, pp. 623–625.

Talmadge, K. *et al.* (1980). 'Bacteria mature preproinsulin to proinsulin' *Proc. Nat. Acad. Sci.* **77**, pp. 3988–3992.

Chapter 8

Growth Hormone

E-PO

Introduction

In the 1920s, the growth-promoting effects of secretions of the pituitary gland were recognized. Certain types of dwarfism were associated with abnormalities of the pituitary function and later giantism was produced in rats when treated with a crude saline extract of the bovine anterior pituitary gland. Removal of the pituitary gland (hypophysectomy) in rats was shown to result in dwarfism, which could be reversed by administration of pituitary extracts.

A sensitive bio-assay for growth promotion of such pituitary extracts was developed in the 1930s. This was based on the growth response of cartilage of hypophysectomised immature rats and the use of such an assay led to the isolation of a pure active protein from bovine pituitary glands — named somatotrophin, or growth hormone. The isolation was based on crude aqueous extracts from which the hormone could be isolated by selective precipitation.

Human hypopituitary dwarfism is usually a hereditary condition based on a deficiency or malfunction of growth hormone (hyposomatotropism). Various attempts have been made to use isolated bovine growth hormone for the treatment of this condition. However, it was not until the 1950s that the failure of human response to such hormone treatment was recognized as a reflection of the exceptional species specificity of growth hormone. Although bovine growth hormone is active in rats, the activity of primate growth hormone is confined to primates. Thus, only human or monkey growth hormone could be used in the treatment of human hypopituitary dwarfism. In 1959, the first successful treatment of human dwarfism was performed by administration of growth hormone derived from isolates of human pituitary glands. Since then national agencies have been established in various countries for the collection of pituitary glands from human cadavers following post-mortem. From such pituitary glands, the growth hormone can be isolated and purified before administration to patients who have been diagnosed as requiring such treatment.

The supply of human growth hormone from this source is limited and insufficient to satisfy the therapeutic demand. The situation is made worse by the increased diagnosis of children lacking growth hormone. Potential beneficiaries include those of below average stature and slow growth but who are not strictly classified as dwarfs. A further complication is the observed contamination of some batches of human pituitaries with virus which causes particular concern as the final product is used for repeated injections in children. For these reasons human growth hormone has had considerable attention with respect to alternative possibilities of large-scale production.

Structure of Growth Hormone

The size of growth hormone was in dispute for many years because of the tendency of the molecule to aggregate — in particular dimers are often found in the blood stream. The molecular weight of the basic molecule is ~20 000 and consists of a single polypeptide chain of ~200 amino acid residues. The tertiary structure is held intact by the presence of disulphide bridges. There are no carbohydrate or lipid groups associated with the polypeptide.

Comparisons of the amino acid sequences of growth hormone between species show significant differences. This results in dissimilar physiochemical properties which necessitate different isolation procedures for the protein. Such sequence differences between species is also reflected in the biological and immunological properties of these isolates. Thus, there is little immunological cross-reactivity between the bovine and human hormone — and non-primate growth hormones are inactive in man.

Various attempts have been made to relate the structure of growth hormone to its biological activity. Reduction of the two disulphide bridges followed by alkylation of the resulting thiol groups produces a derivative that retains biological activity. Also, tryptic digestion of the hormone releases a small peptide of 39 amino acid residues that retains some of the growth-promoting activity. These studies show that the full 3-dimensional tertiary structure of the molecule may not be required for activity.

Physiological Production of Growth Hormone

Growth hormone is released by the somatotropic cells found in the anterior pituitary glands. These somatotrophs are the predominant cells of the anterior pituitary and growth hormone can make up ~40% of the total hormone content of this gland. Human growth hormone (hGH) release is under regulatory control by factors secreted by the hypothalamus. The two most important hypothalamic secreted factors to be isolated are somatostatin which inhibits growth hormone release and growth hormone releasing factor (GHRF) which stimulates release. The latter was isolated as a single molecule in 1982 and was found to consist of a 44 amino acid peptide which has a very short half life ($\sim$1 min) *in vivo*.

The concentration of growth hormone in the blood stream varies from a basal level of $\sim$2 ng/ml up to $\sim$30 ng/ml upon stimulated release. Such higher levels are normally found in children and adolescents. The primary effect of the hormone is stimulation of cartilage growth although growth promotion is also found of soft tissue and muscle.

Growth hormone receptors are found in plasma membrane particularly in liver cells. Interaction with such receptors would appear to stimulate the release of several growth factors. The most thoroughly characterized of these are a family of molecules called somatomedins — small peptides (molecular weight 8–9000) which mediate most of the growth activity of the cartilage. They were originally called sulphation factors because of their ability to promote sulphate incorporation into growing cartilage.

The interaction of the several peptides mentioned above (Fig. 8.1) is typical of several known hormone cascades in which a particular biological effect such as growth promotion is stiulated by a sequence of hormones, each member of the sequence stimulating the release of the next.

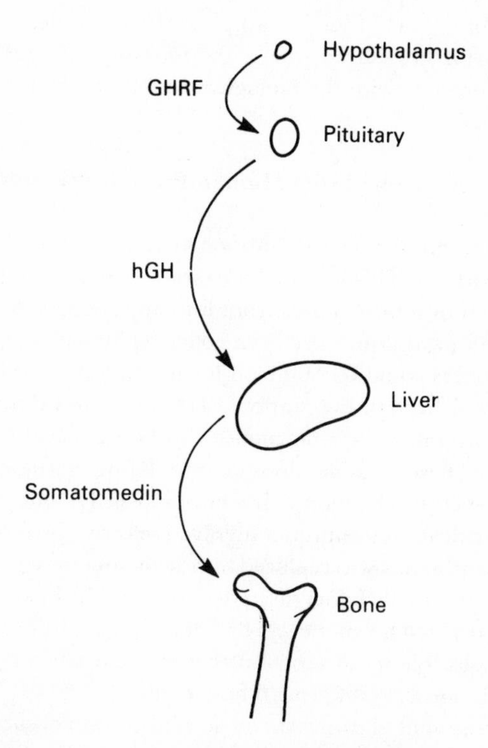

Fig. 8.1 The growth hormone cascade.

Growth Abnormality and Diagnosis by Hormone Assay

Growth abnormalities can result from the malfunction of any of the associated molecular interactions. These can include the defective release of — growth hormone releasing factor, growth hormone or somatomedins. Alternatively, it can result from defective cell membrane receptors required to mediate the activity of any of these hormones. It therefore follows that only a certain number of cases of dwarfism can be treated by administration of growth hormone, i.e. hypopituitary dwarfism caused directly by insufficient growth hormone. Diagnosis of this condition is required by assay of hormone levels in the blood.

Although non-primate growth hormone is inactive in man, it is fortunate for assay purposes that human growth hormone is active in non-primates. Thus, the early assay of the hormone involved measurements of the increase in body weight of hypophysectomised rats following hormone injection. Such a body weight test was made more sensitive by the development of a genetic strain of hypopituitary dwarf mice. An alternative assay which has been used extensively involves measurement of the increase in width of epiphyseal cartilage of the tibia of rats.

Although such a cartilage assay is much faster than those involving overall body weight, it still has the disadvantages of high cost, low sensitivity and variability. The development of an immunoassay using monoclonal antibodies (see Chapter 6) has solved many of these problems and allows the routine measurement of growth hormone in large numbers of serum samples. However, despite the sensitivity of the immunoassay for growth hormone, the immunogenic response of a sample does not always correlate with the biological activity of the hormone.

Extraction of the Hormone From Human Pituitary Glands

In 1959 the value of human growth hormone was recognised for the treatment of hypopituitary dwarfism. This initiated the establishment of national programmes for the collection of pituitary glands from human cadavers during post-mortem. Since then, human pituitaries have been collected by pathologists so that growth hormone-rich extracts could be obtained for use in human therapy. In the USA this was established through the National Hormone and Pituitary Program. As well as growth hormone, other hormones can be extracted from such banks of pituitary glands. These include thyroid-stimulating hormone (TSH), follicle-stimulating hormone (FSH), luteinizing hormone (LH) and prolactin.

The earliest purification techniques involved selective precipitation of an active protein. Subsequently, this was replaced by a sequential series of chromatographic or electrophoretic steps which increased the purity of the final product. Typically, about 10 mg of extracted hormone can be obtained per pituitary gland. This must be related to the possible requirement of 1 g per year per recipient — treatment being given continuously until growth has ceased.

Therapeutic treatment of dwarfism by administration of such human extracted and purified hormone was continued over a period of 25 years from its start in

1959. However, in 1984 a major problem arose following the death of a patient with Creutzfeldt-Jakob disease. This is a relatively rare viral infection of the brain, which in this case was transmitted by administration of the growth hormone. Subsequent cases of hormone-associated viral infection resulted in a decision to suspend all further hormone treatment.

The total suspension of growth hormone treatment was justified by the risk of infection of this potentially fatal disease being transmitted by treatment of a condition — dwarfism — which is not a life threatening condition. Thus the risk-to-benefit ratio was considered too high to continue use of the hormone therapy. Such risks must always be considered when a product is extracted from biological tissue of mixed and unknown origin. Similar risks are associated with the extraction of clotting factors from donated blood plasma (see Chapter 10).

In the case of pituitary-derived human growth hormone, the viral contamination can now be eliminated by treatment of the pituitary extract with 0.1M NaOH or filtration through a 25 nm filter. However, the incidences of Creutzfeldt-Jakob disease have accelerated the use of growth hormone produced from genetically engineered cells. Thus in 1985, biosynthetic human growth hormone became the second recombinant DNA product to be approved for human use.

Growth Hormone from Genetically Engineered Cells

The expression of growth hormone by bacteria was considered by Genentech and reported by Goeddel *et al.* in 1979. The primary translation product of growth hormone mRNA in mammalian cells is a precursor molecule consisting of a signal peptide which is subsequently removed before secretion. The proteolytic cleavage of this precursor molecule which takes place in eucaryotic cells is unlikely to occur in procaryotes because of the lack of the appropriate processing enzymes. Thus expression of the whole human gene in bacteria would result in the synthesis of a pre-hormone containing an extra 26 amino acids. This problem of post translational processing as well as the large size of the growth hormone molecule were important considerations in the development of a strategy for producing a suitable genetically engineered bacterium.

A successful strategy developed for the production of growth hormone from a genetically engineered strain of bacterium was as follows (Fig. 8.2):

(1) Growth hormone mRNA was extracted from human pituitary cells
(2) The corresponding complementary DNA was prepared from this by use of the enzyme, reverse transcriptase. This cDNA contained the nucleotide sequences for growth hormone as well as the precursor fragment on the 3′ end. The sequence of the cDNA and positions of the restriction endonuclease sites were determined. Of particular interest were the HaeIII endonuclease sites which were present in the signal sequence and one in the coding sequence at amino acids 23–24 of the growth hormone
(3) The double stranded cDNA was treated with HaeIII which cleaved the

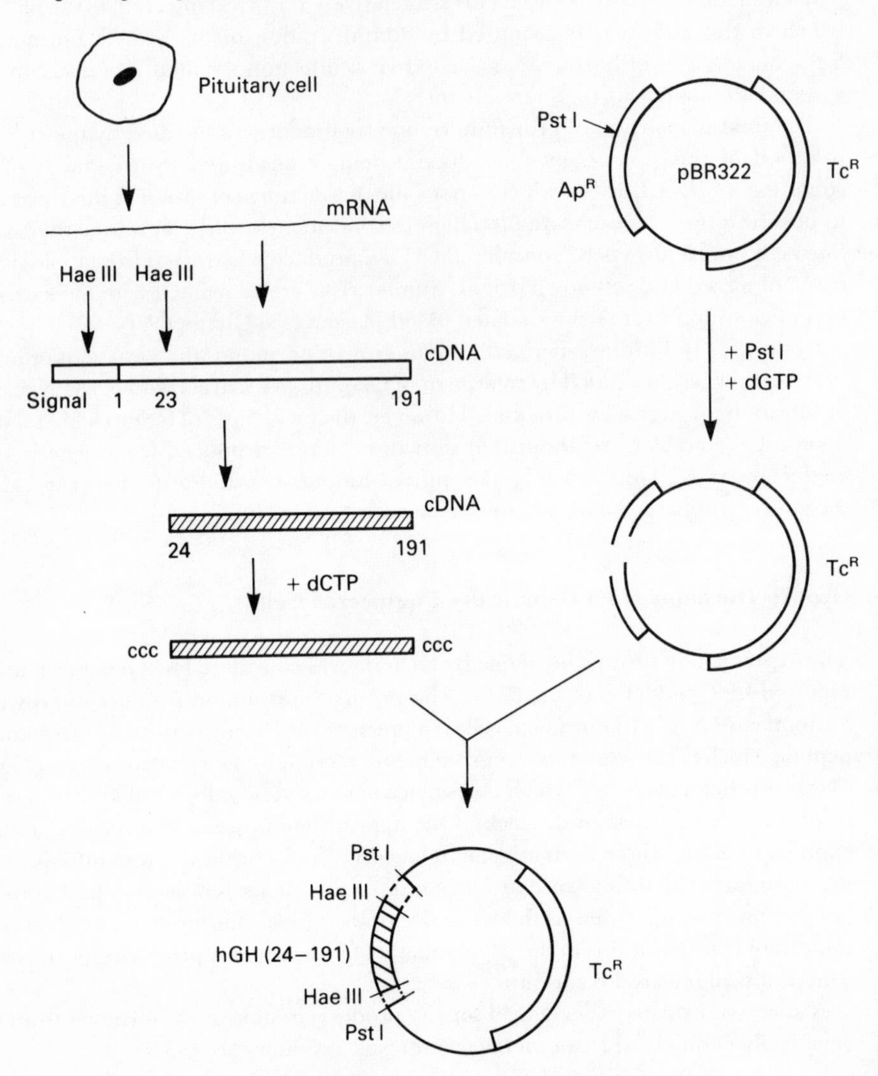

Fig. 8.2 Assembly of hGH gene.

(1) Cloning of large fragment

(2) Cloning of small fragment

(3) Ligation of 2 fragments

(2) *Cloning of small fragment*

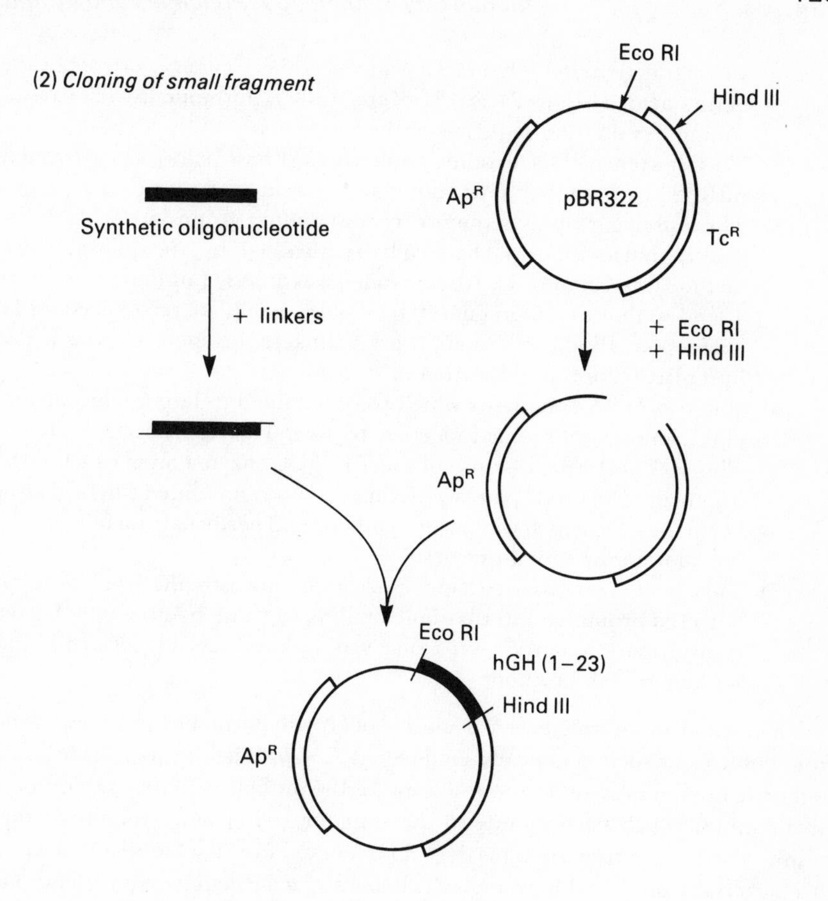

(3) *Ligation of 2 fragments*

Gene fragments obtained from cloning

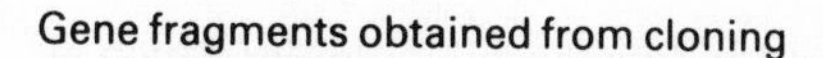

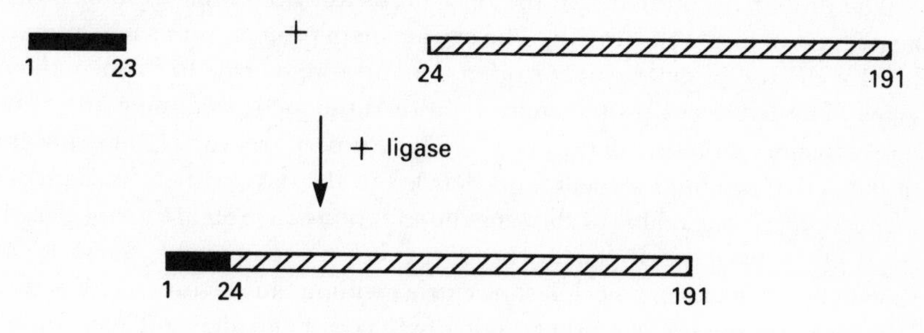

Complete hGH gene

signal fragment but left intact a protein coding sequence corresponding to amino acid residues 24 to 191 of the growth hormone as well as a non-coding termination sequence at the 5′ end

(4) This shortened cDNA coding sequence (551 base pairs) was inserted into pBR322 by homopolymer tailing and transformed into *E. coli*

(5) The missing coding sequence corresponding to amino acids 1–23 of growth hormone was chemically synthesized by the phosphotriester method (see Chapter 7). Also included was an ATG initiator codon at the 3′ end so that the fragment (40 base pairs) would be recognized for later expression. By the use of appropriate linkers this molecule was inserted into pBR322 for amplification in *E. coli*

(6) The two DNA fragments which were obtained in larger concentrations after cloning were joined together by their HaeIII restriction site ends. This was performed by use of the T4 ligase enzyme after which a DNA fragment of 591 base pairs was isolated. This represented a hybrid growth hormone gene which was partly synthetic and partly natural but which did not contain the signal precursor

(7) This gene was inserted into a plasmid downstream from a suitable bacterial promoter and ribosome binding site. After subsequent bacterial transformation, an *E. coli* clone was isolated capable of synthesizing human growth hormone

A high yield amounting to ~ 2.4 mg l^{-1} of growth hormone has been produced from cultures of such genetically engineered *E. coli*. The product is identical to authentic human growth hormone except for the presence of an extra amino acid (methionine) which corresponds to the initiator codon and remains at the N terminal end of the protein. This does not seem to affect the biological activity of this bacterially produced hormone which has a specific activity equivalent to that derived from human pituitary cells. However, the antigenic activity of such methionine derivatives of growth hormone may be greater and could cause antibody formation in treated patients. The importance of this is yet to be established.

The protein accumulates in the *E. coli* cells as cytoplasmic granules which are aggregates of insoluble protein. The extraction of protein from such granules may involve the use of detergents or other denaturating agents to dissolve the aggregates. The extracted protein may require subsequent treatment to restore the native tertiary structure of the protein. These procedures can result in lower yields of the active product. Recent work has led to the development of a procaryotic expression vector capable of allowing the secretion of a protein into the periplasmic space of the bacteria. In such experiments it has been possible to isolate human growth hormone from a periplasmic extract without any significant denaturation. Furthermore, the use of a procaryotic signal vector can allow the formation of the human growth hormone with the correct N terminal.

The growth hormone synthesized by genetically engineered bacteria gained approval in 1985 and is now being produced on a large scale. This was only the second protein produced by recombinant DNA technology to be approved for human therapeutic use. Undoubtedly, the rapid approval of this product was a reflection of the need to find a replacement for the virally contaminated product from pituitary glands.

Alternative Methods for Growth Hormone Synthesis

In 1970 human growth hormone was first produced by chemical synthesis. Despite the demonstration of biological activity in this molecule, this approach has not been pursued for commercial production. The size of the molecule is prohibitively large to allow a large-scale production method based on such synthesis.

The production of growth hormone from pituitary cell cultures has been studied particularly with a rat pituitary cell line which has shown a capability for continuous hormone secretion *in vitro*. Such secretion can be increased $\times 7$ by the presence of hydrocortisone in the medium. Similar *in vitro* stimulation can be shown by thyroid hormones or by glucocorticoids. Despite this work, it has not been considered viable to develop large-scale production from such cells. Human pituitary cells would have to be used and the yield of hormone from such cells is unlikely to be sufficiently high to allow commercial production.

However, the use of genetically engineered animal cells for growth hormone production is a viable proposition which is being developed by several commercial companies. The use of such cells could allow appropriate post-translational modification and the secretion of the protein into the growth medium. This could ease the problems of purification and result in an authentic human growth hormone which would not produce an immunogenic response.

The Use of Growth Hormone in Animals

Bovine growth hormone injected into cattle will cause growth stimulation which can have advantages in farming and meat production. Such growth stimulation can be produced by the injection of any species specific hormone into the appropriate animal.

In 1982 this approach was taken one stage further by the insertion of the growth hormone gene into the fertilised egg of a mouse by micro-injection. The gene together with its associated promoter was incorporated into the chromosomal DNA and the resulting mice progeny was capable of high levels of growth hormone synthesis. These transgenic mice were capable of growth to significantly ($\times 2$–$\times 3$) greater size than the controls. The incorporation of the gene into the chromosome of the embryonic cells enabled the high producing growth hormone gene to be preserved and expressed through subsequent generations.

The promoter used was taken from a metallothionein gene which is induced by heavy metals. As a result growth hormone synthesis in the mice could be induced by the administration of heavy metals. High levels of growth hormone synthesis was not confined to the pituitary gland but was detected in various tissues not normally associated with its synthesis.

This work has clear implications in the possibilities of obtaining strains of transgenic animals with desired characteristics for farming. Also, the possibilities of extracting growth hormone or other high value products from such animals may be considered.

Conclusion

Growth hormone is a natural product of differentiated pituitary cells *in vivo*. Such cells have not been used extensively for large-scale production *in vitro* because of the difficulties of growth and low productivity levels. The relatively small size of the growth hormone molecule has made the manipulation of its gene relatively easy and production levels from the resulting isolated recombinant *E. coli* have been high. The recombinant product has been licensed and used extensively since 1985.

However two problems remain—the recombinant product from *E. coli* has an extra methionine residue which may induce an unwanted immunogenic response and the product is accumulated intracellularly, thus necessitating stringent purification techniques from the bacterial lysate. Alternative processes include chemical synthesis, which is difficult to scale-up as a viable operation because of the size of the molecule. Large-scale production from cultured pituitary cells is unlikely to be acceptable because of the low level of product secretion even with the use of specific inducers.

Future production processes may involve the use of genetically engineered animal cells—particularly Vero and murine cells—which are being actively developed by several commercial companies. Such cells have the advantage of high level secretion of an authentic human product. Also, extraction from the cell medium can be a simpler and cheaper process compared to extraction from cell lysates.

Summary

Growth hormone is a non-glycosylated protein of 20 000 molecular weight secreted by cells of the anterior pituitary glands. Its secretion is under regulatory control by hormones secreted by the hypothalamus. Growth hormone levels are generally high in children and adolescents when tissue (and particularly cartilage growth) is stimulated through various intermediary growth factors.

Malfunction of any of the hormones or factors involved in the cascade of interactions can lead to abnormal growth patterns. One such abnormality — human hypopituitary dwarfism — can often be attributed to an insufficient production of growth hormone. Therapeutic treatment of this condition can be affected by a course of injections of growth hormone which has to be of human origin because of the extreme species specificity of the biological activity of the hormone.

The value of growth hormone for the treatment of human hypopituitary dwarfism was recognized in the 1950s and prompted a need for its large-scale production or extraction. For 25 years the sole source of human growth hormone was from pituitary gland extracts of human cadavers. National collections of human pituitary glands enabled large-scale extraction and purification of the hormone. However, the success of its therapeutic use resulted in an increased demand for the hormone which could not be easily met from pituitary extraction. A further problem resulted from cases of the fatal Creutzfeldt-Jakob disease which were induced by treatment of virally infected pituitary extracts. As a result, the use of pituitary extracts for therapy was suspended and alternative sources of the hormone considered.

The therapeutic use of human growth hormone from recombinant bacteria was accepted in 1985. Production of the genetically engineered bacteria capable of growth hormone synthesis required a clonal strategy which could take account of the post-translational cleavage of a precursor molecule, which normally occurs in secreting pituitary cells. This problem was solved by insertion of a modified gene into a plasmid which could be expressed in *E. coli*. Such a recombinant product has been used extensively since 1985. However, future developments are likely to result in the increased use of genetically engineered animal cells for large-scale production.

The experimental techniques of gene micro-injection into embryonic cells have been applied to the insertion of the growth hormone gene into the germ line of a mouse strain. The use of these techniques for developing novel strains of transgenic animals may have considerable future applications.

General Reading

Cox, R.P. and Day, D.G. (1981). 'Regulation of glycopeptide hormone synthesis in cell culture', *Adv. Cell Cult.* **1**, pp. 15–65.

Glasbrenner, K. (1986). 'Technology spurt resolves growth hormone problem, ends shortage', *J.A.M.A.* **255**, pp. 581–587.

Katinger, H.W.D. and Bliem, R. (1983). 'Production of enzymes and hormones by mammalian cell culture', *Adv. Biotech. Processes* **2**, pp. 61–95.

Tashjian, A.H.Jr (1969). 'Animal cell cultures as a source of hormones', *Biotech. Bioeng.* **11**, pp. 109–126.

Wallis, M., Howell, S.L. and Taylor, K.W. (1985). *Biochemistry of the Polypeptide Hormones*. Chichester, J. Wiley & Sons.

Specific Reading

Bewley, T.A. *et al.* (1969). 'Human pituitary growth hormone. Physiological investigation of the native and reduced-alkylated protein', *Biochemistry* **8**, pp. 4701–4708.

Goeddel, D.V. *et al.* (1979). 'Direct expression in *E. coli* of a DNA sequence coding for human growth hormone', *Nature* **281**, pp. 544–548.

Gray, G.L. *et al.* (1985). 'Periplasmic production of correctly processed human growth hormone in *E. coli*: natural and bacterial signal sequences are interchangeable', *Gene* **39**, pp. 247–254.

Hsuing, H.M. *et al.* (1986). 'High level expression efficiency secretion and folding of human growth hormone in *E. coli*', *Biotechnology* **4**, pp. 991–995.

Martial, J.A. *et al.* (1979). 'Human growth hormone: complementary DNA cloning and expression in bacteria', *Science* **205**, pp. 602–606.

Olson, K.C. *et al.* (1981). 'Purified human growth hormone from *E. coli* is biologically active', *Nature* **293**, pp. 408–411.

Palmiter, R.D. (1982). 'Dramatic growth of mice that develop from eggs microinjected with metallothionein-growth hormone fusion genes', *Nature* **300**, pp. 611–615.

Schoner, R.G. *et al.* (1985). 'Isolation and purification of protein granules from E. coli cells overproducing bovine growth hormone', *Biotechnology* **3**, pp. 151–154.

Seeburg, P.H. *et al.* (1977). 'Nucleotide sequence and amplification in bacteria of structural gene for rat growth hormone', *Nature* **270**, pp. 486–494.

Seeburg, P.H. *et al.* (1978). 'Synthesis of growth hormone in bacteria', *Nature* **276**, pp. 795–798.

Chapter 9

Plasminogen Activators

Introduction

One of the major causes of death in the Western world is thrombosis, which is a condition associated with a blood circulatory blockage caused by the deposition of a fibrin clot or thrombus. This can lead to one of a number of potentially fatal conditions, e.g. myocardial infarction, pulmonary embolism, arterial thrombo-embolism. Early diagnosis of a thrombosis enables treatment by surgical removal of the thrombus or by the use of one of a group of enzymes capable of dissolving the fibrin clot. Such a group of enzymes are the plasminogen activators which are capable of initiating the fibrinolytic process *in vivo* and causing breakdown of the fibrin.

In 1933 an extracellular protein released from various strains of *Streptococci* were shown to be capable of dissolving blood clots *in vitro*. Later it was demonstrated that a purified sample of this enzyme — streptokinase — was an activator of human plasminogen which can initiate fibrin breakdown *in vivo*. By the 1950s the full therapeutic potential of streptokinase was realized and high purity samples (>95%) were made available for intravenous use in man for the treatment of thrombosis. Since then the enzyme has been used successfully for the treatment of various forms of thrombosis. However, despite its success and continued use there are a number of problems associated with streptokinase. Pyrogenic and antigenic reactions are common, and there is an increased danger of haemorrhaging caused by a reduced clotting ability of the blood. The pyrogenic reactions, which were particularly observed in the early days of the use of the enzyme, were reduced by improved purification techniques developed in the 1960s. However, the antigenic response is a reflection of the non-human origin of the enzyme.

A similar enzyme of human origin and capable of activating fibrinolysis was discovered in the 1950s. This enzyme could be extracted and purified from human urine and was termed urokinase. The purified enzyme could be used for similar therapeutic treatment as streptokinase but without the disadvantages of pyrogenic, allergic or antigenic reactions. Urokinase is commercially available as a purified enzyme extracted from urine. However, its large scale use has been hampered by the difficulties of handling and extraction from the large quantities of urine required as a source.

In the 1970s, a third type of enzyme capable of initiating fibrinolysis was discovered. This is termed tissue-type plasminogen activator (or t-PA) and is found to occur naturally in the human bloodstream as a secreted product from cells of a variety of tissues. t-PA has a number of advantages for therapeutic use in the treatment of thrombosis. It can be derived from human cells and therefore like urokinase does not cause any undesirable immunogenic reactions. However, t-PA is far more specific than urokinase in its binding to a fibrin clot. This localizes fibrinolysis and so limits the problem of haemorrhaging which may be caused by a reduction in the overall clotting ability of the blood. Thus t-PA is more effective than either streptokinase or urokinase in its activation of fibrinolysis at the site of a clot. As the therapeutic advantages of t-PA are clear, the challenge for cell technology is to devise a process for the large-scale production of the enzyme.

The Physiological Process of Fibrinolysis

Fibrin is an insoluble protein formed during blood coagulation and involved in the early phase of tissue repair. The process is initiated through a cascade of clotting factors (see Chapter 10) which finally leads to the conversion of soluble fibrinogen to a solid deposit of fibrin. This conversion normally occurs at the site of injured tissue where a solid fibrinous deposit can provide a barrier against loss of blood, i.e. haemorrhage. Fibrin formation involves the polymerization of the fibrino monomer released by the proteolytic action of thrombin on fibrinogen. Repair tissue is later formed by fibroblast cells migrating into the fibrin layer.

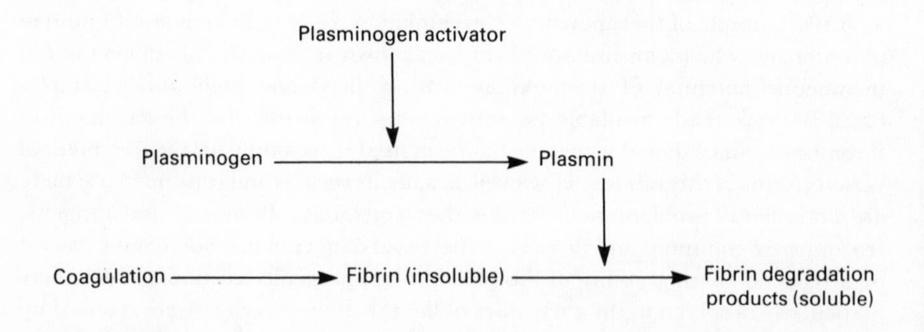

Fig. 9.1 Fibrinolysis.

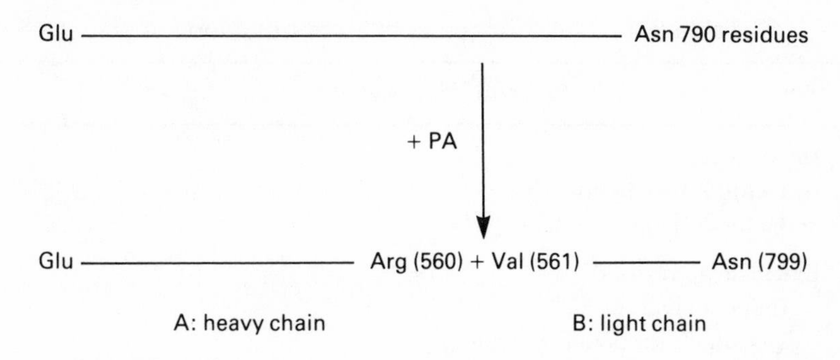

Fig. 9.2 Activation of plasminogen by proteolytic cleavage.

Fibrinolysis is the the natural counterpart to this clotting process and involves the activation of a cascade of factors culminating in the dissolution of fibrin by the proteolytic activity of plasmin (Fig. 9.1). As in the clotting process, the fibrinolytic factors are activated in a series of proteolytic cleavage reactions. The plasminogen activators in the blood stream, can convert plasminogen to plasmin, which in turn leads to the proteolytic breakdown of fibrin.

Plasminogen is a glycoprotein with a molecular weight of 92 000 and is known to be synthesized in the liver. Activation of the plasminogen molecule is known to involve a cleavage into two peptide chains by the action of a plasminogen activator on an Arg-Val linkage (Fig. 9.2). One of the chains (the A chain) may be further fragmented before forming an active complex with the B chain by disulphide bridges.

There are several types of plasminogen activators which may be classified according to their origin (Table 9.1) or according to their structure and properties (see below). The intrinsic activators are those normally present in the blood stream and can be sub-classified into those dependent or independent of the blood clotting factor XII. These intrinsic activators are normally present in an inactive form and their activation depends upon the presence of plasmin, thus making the process autocatalytic.

Plasminogen activators can be secreted by cells of a range of mammalian tissues and these are known as the extrinsic activators. The vascular endothelium is responsible for secreting such factors into the blood stream. The secretion of such extrinsic activators can be stimulated by a number of substances including thrombin and histamine and also by exercise. Stimulation can cause the blood concentration of the activators to rise from a base level of ~ 1 ng/ml to 100 ng/ml. However the higher concentration levels do not last very long, as clearance by the liver is rapid — the typical half-life of t-PA is 2 to 4 min.

The activators found in the urine are secreted by kidney cells and are thought to be physiologically important in preventing fibrin blockage of the urinary tract. The exogenous activators are those derived from sources other than mammalian cells but have been shown capable of activating fibrinolysis *in vitro* or *in vivo*. Of these, streptokinase in particular has been important in therapeutic use.

Table 9.1 Plasminogen activators of various origin

(a) Physiological
(i) Intrinsic activators
— factor XII-dependent
— factor XII-independent
(ii) Extrinsic activators
— tissue activator
— vascular endothelial activator
(iii) Urinary activator
(b) Non-physiological
(iv) Bacterial
— Streptokinase
— Staphylokinase
(v) Chemical
— chloroform
— glycerol

Under normal physiological conditions the clotting factors are in a homeostatic balance with the fibrinolytic factors which are only activated when tissue is damaged. An inbalance of these processes can cause either of two dangerous pathological changes. A reduced ability for blood coagulation or an enhanced ability for fibrinolysis can result in excessive haemorrhaging following tissue damage. Alternatively, an enhanced ability for coagulation or a reduced ability for fibrinolysis can lead to the unwanted formation of a fibrin clot resulting in a thrombotic condition. If the inside of blood vessels become damaged or are subject to deposits of for example cholesterol, there may be a tendency for the blood to clot causing the formation of a thrombus. The clot may be removed by the natural physiological activation of fibrinolysis. However, the therapeutic administration of plasminogen activators is often necessary to enhance this activation and promote rapid thrombolysis.

Structure of the Plasminogen Activators

There are two main immunologically distinct types of plasminogen activator found in mammals — urokinase (u-PA) and tissue-type (t-PA). Although urokinase was named after its original discovery in urine, it has subsequently been found in numerous tissues particularly kidney and in blood plasma. There are two main types of urokinase which have been isolated from urine and kidney cell cultures.

They are characterized as glycoproteins with mean molecular weights of 30 000 (low molecular weight urokinase) and 55 000 (high molecular weight urokinase). A zymogen known as pro-urokinase has also been isolated.

Pro-urokinase is a single peptide chain of molecular weight 55 000 and can be converted to the high molecular weight urokinase by the action of plasmin, which causes proteolytic cleavage to a two chain structure without the loss of overall size. The low molecular weight urokinase is formed by fragmentation of a peptide from the high molecular weight form.

Tissue-type plasminogen activator (t-PA) is immunologically distinct from urokinase although a significant sequence homology has been shown. t-PA has been characterized as a glycoprotein of slightly larger molecular weight — 56–83 000. The exact molecular size of t-PA may differ between the cell type of origin and can depend upon the variable degree of glycosylation and proteolytic degradation. In some cells a single polypeptide chain t-PA has been shown to be broken down by proteolysis to a lower molecular weight two-chain protein.

The t-PA secreted by Bowes melanoma cells has been well characterized as a 547 amino acid residue polypeptide chain with 4 potential glycosylation sites at asparagine positions. The extent of glycosylation is typically ~7% of the total molecular weight. Its polypeptide structure can be divided into several functional domains (Fig. 9.3). Each structural domain is encoded by separate exons of the

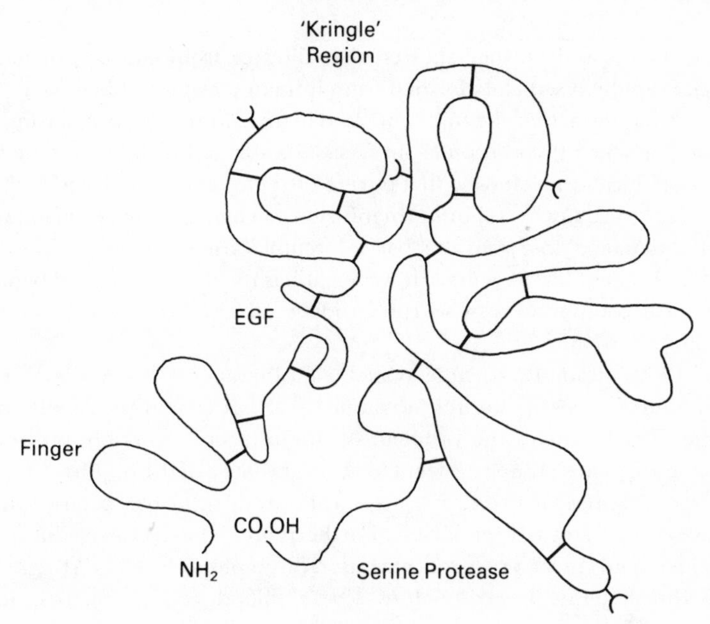

The various domains of the t–PA molecule are shown in relation to its secondary structure, which is held together by disulphide bridges. The four possible glycosylation sites are indicated by a Y

Fig. 9.3 Proposed structure of tissue-type plasminogen activator. (From Klausner, 1986)

t-PA gene which is in agreement with the idea that exons represent independent genetic building blocks. The catalytically active protease domain has been demonstrated in a region of 244 amino acid residues at the C terminal end and this shows the greatest homology to other serine proteases, e.g. trypsin, chymotrypsin and plasmin. The mid-part of the molecule is a domain of 80 amino acid residues called the 'kringle' region characterized by a looped structure held together by three disulphide bridges. There are two kringle structures in t-PA which have homologous sequences to one kringle structure in urokinase and five in plasminogen. An adjacent domain of unknown function has homology with sequences found in epidermal growth factor and transforming growth factor and the N terminal domain called the 'finger' region has homologous sequences to a domain in fibronectin which binds to fibrin.

The kringle and the finger domains have been implicated in fibrin binding. This is important for the therapeutic use of t-PA so as to localize the activation of fibrinolysis to the site of a fibrin clot. The urokinase molecule also has a kringle domain but the susceptibility of the molecule to proteolytic degradation *in vivo* results in its inability to specifically bind to fibrin.

Therapeutic Use of Plasminogen Activators

In 1933, Tillett and Garner showed that filtrates from cultures of haemolytic *Streptococci* rapidly lysed clots formed from human plasma or fibrinogen. Streptokinase which is the active enzyme contained in this filtrate can form a complex with plasminogen which then becomes more susceptable to breakdown to plasmin. In the 1950s, clinical trials showed that purified streptokinase could lead to the lysis of thrombi *in vivo* usually by a short term infusion of a high priming dose followed by a lower maintenance dose. On this basis, streptokinase has been widely used as a thrombolytic agent for 30 years. However, it has the disadvantage of being a non-physiological compound and so can produce adverse immunological reactions during human treatment.

In the 1960s urokinase became available for therapeutic use and as it is derived from human sources it has the advantages of being non-pyrogenic and non-antigenic. This removes the problem of immunogenic reaction and makes the determination of dose much easier. However, because of the required purification from large volumes of urine, its large-scale production has been prohibitively expensive compared to streptokinase. Furthermore, it has a low specific binding to fibrin which tends to lead to a concentration throughout the blood stream that can lead to the systemic breakdown of several blood protein factors, including fibrinogen. This lowers the clotting ability of the blood and may result in haemorrhaging.

Whereas streptokinase (derived from microbial cultures) and urokinase (derived from urine) have been used therapeutically for a number of years, the effect of tissue-type plasminogen activator was only reported in 1980 following sufficient production from a melanoma cell line to allow a limited study of its therapeutic application. There are a number of advantages of the use of t-PA compared to the other activators:

(1) It is a natural component of blood with a typical base plasma level of 1 ng/ml which can rise to ~100 ng/ml under certain conditions, e.g. during exercise. The therapeutic level of t-PA can be kept within the physiological range
(2) It has a high affinity for fibrin and thus the localized concentration of t-PA around a fibrin clot is high
(3) Its catalytic activity is stimulated by fibrin. Thus the concentrations of t-PA that are effective in thrombolysis do not have significant activity in the circulating blood away from the clot. This limits the systemic degradation of blood proteins which can cause problems of excessive bleeding
(4) Its clearance from the circulating blood is rapid: half-life is ~2–4 min. This has two consequences — it increases the doses required for therapy but it means that therapy can be well controlled and can be terminated instantaneously.

Assay Methods

The fibrin plate method has been used extensively to demonstrate the fibrinolytic activity of tissues, biological fluids and purified plasminogen activators. In this method a fibrin layer is prepared on a surface such as the bottom of a Petri dish by mixing solutions of bovine thrombin and fibrinogen. The fibrinogen preparation used contains residual concentrations of plasminogen. Fibrinolytically active material placed on top of this fibrin layer will activate this plasminogen which will produce a zone of visible lysis of fibrin, the area of which is proportional to plasminogen activation. The assay gives variable results because of the varying concentrations of fibrinogen and plasminogen contained in different plasma extracts used in formation of the fibrin layer. To allow for this, the fibrinolytic activity of a soluble sample is measured at a range of dilutions and compared with a standard urokinase sample which can be provided by the international Committee of Thrombolytic Agents (CTA) who have established a generally recognized standard of activity as a CTA urokinase unit.

This fibrin plate method has been used to show fibrinolytic activity in various tissue explants. A large variation in activity is shown between tissue types with particularly high activity been shown in vascularised connective, uterine and kidney tissue.

A synthetic chromogenic plasmin substrate is available to provide a more rapid assay than the fibrin plate method and can be used for extracted samples. This is based upon the production of plasmin by an activator which in turn can cleave Val-Leu-Lys-p-nitroanilide into p-nitroaniline, the concentration of which can be measured colorimetrically by its absorbance at 405 nm (Fig. 9.4). Alternatively

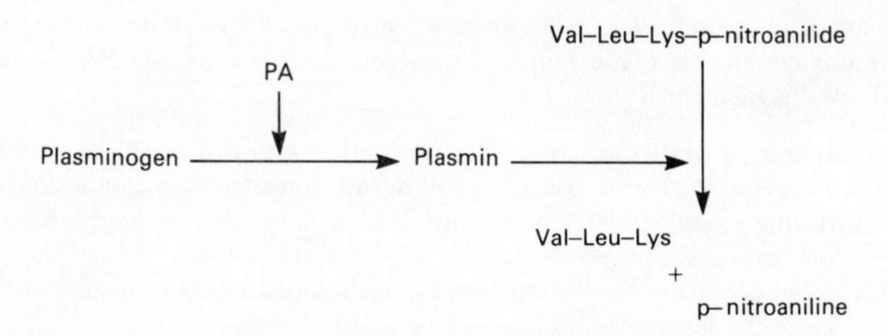

Fig. 9.4 A simple chromogenic assay for extracts of plasminogen activator.

such esterolytic activity can be measured by the action of the enzyme against the synthetic substrate, acetyl-L-lysine methyl ester (ALME), the products from which can also be measured colorimetrically. These assays are suitable for use in a multi-titre plate.

Antibodies have been raised to the various plasminogen activators and rapid ELISA assays are now available (see Chapter 6) and are suitable for the extracted or purified samples.

Large-Scale Production of Physiological Plasminogen Activators

Urokinase

The urokinase content of urine arises from secretions of the epithelial cells of the urinary tract. However, the major problem of urokinase extraction from urine is that the enzyme is present at low and variable levels. This necessitates an initial concentration step before purification. Two methods have been extensively used:

(1) The pH of cooled urine is lowered to 4.3, and this results in a protein precipitate containing ~90% of the original urokinase
(2) The urine can be agitated, shaken or aerated to cause foaming. As a result of this, a high proportion of the urokinase accumulates in the foam

These methods can lead to ~ ×20 concentration of urokinase and must be followed by a series of classical protein purifications steps which may consist of ammonium sulphate precipitation, ion exchange chromatography or gel filtration. Alternatively, an affinity column which contains a competitive inhibitor to the enzyme activity can be used. Such a compound is α-N-Benzylsulphonyl-p-amino-phenylalanine which can be bound covalently to Sepharose gel and loaded into a column. The binding of urokinase to this compound is highly specific and other unwanted proteins can be washed through the column. Final elution of the pure urokinase can be accomplished by use of a second soluble competitive inhibitor or more simply by a high salt solution. This method of affinity chromatography is a simple method and the column can be used several times for different extracts.

By these extraction and purification methods the typical yield of urokinase from a litre of urine is 6000 units. This must be compared with the recommended course for thrombolytic therapy of 4000 units kg^{-1} h^{-1} up to 72 h. This calculates to a requirement of 21×10^6 units per course of therapy for an average patient. The quantity of urine required for the extraction of this quantity of urokinase is 3500 l which might be equated as the urine output of 12 men over 6 months. This calculation clearly illustrates the problem of this extraction procedure as a method for the large-scale supply of urokinase and accounts for its prohibitive cost.

The urokinase isolated from urine originates from the lining of the kidney tubules and so cells derived from this tissue can be used in culture to generate the same urokinase enzymes. This approach was made possible in the early 1970s when primary cultures from kidney tissue of a human foetus were shown to be able to produce urokinase. Although low yields may occur from such cultures, some reports have shown that by stimulation these cells can yield up to 80 0000 units/ml of medium following a prolonged period of incubation of a confluent culture.

Such foetal kidney cells have also been found suitable as a source for extracting mRNA for gene cloning of urokinase. In this process, the required fractions of the polyA containing RNA from such cells were assayed for urokinase synthesis in a cell-free protein synthesis mixture derived from rabbit reticulocytes. The selected fractions were then transcribed to cDNA by reverse transcriptase and inserted into the PstI site of the plasmid pBR322 (see Chapter 3). Recombinant DNA clones of *E. coli* were subsequently selected for urokinase synthesis. However, although it is feasible to produce urokinase from such clones, they are now unlikely to be used on a large scale because of the generally recognised superior qualities of t-PA as a therapeutic agent.

Tissue-Type Plasminogen Activator

In 1979 t-PA was first purified from uterine tissue which has the highest concentration of any mammalian tissue type. From 5 kg of such tissue, 1.2 mg of t-PA can be extracted. This can be compared with extraction from blood plasma, 1000 l of which would be required for 1 mg of t-PA at the basal concentration, or 44 l if t-PA could be induced prior to extraction. The difficulty of such methods for large-scale production is that sufficient human starting material would not be available to satisfy the present or projected demand for t-PA.

Production from cell cultures has been examined with the use of three classes of cells. Typical t-PA production levels achievable with each cell type are given in brackets:

(1) Immortal cell lines ($\sim$30 units/10^6 cells/day). One particular cell line — the Bowes melanoma line — has been well studied for the production of t-PA. These cells grow as a monolayer culture to high densities and release t-PA into a medium which can be serum-free to allow ease of t-PA extraction. Such cells have been used extensively in laboratory experiments to study the structure and synthesis of t-PA. However, the disadvantage of using such cultures for large-scale production is that

transforming growth factors of a similar molecular size to t-PA can also be released into the medium. This means that such a source of t-PA is unlikely to be acceptable to regulatory authorities for therapeutic use requiring repeated human injections

(2) Primary cell cultures (60–80 units/10^6 cells/day). Primary cells have been found capable of synthesis of both u-PA and t-PA. The efficiency of production depends upon the source of the cells, foetal kidney cells being suitable for u-PA synthesis and vascular endothelial cells particularly suitable for t-PA. Such cultures are generally based on a mixture of cell types, and only some of these might be capable of t-PA synthesis. The cultures can only be passaged for a limited number of generations before product synthesis declines

(3) Human diploid fibroblasts (100–300 units/10^6 cells/day). Some human embryonic lung fibroblasts, notably cell lines IMR-90, WI-38 and HEL129 have been successfully used for t-PA production. The advantage of using such cells is that there is no concomitant production of transforming growth factors. The fibroblasts can be adapted to high t-PA production rates by careful control of the culture conditions. Such adaptation involves a transfer of cells to permissive media conditions in which the cells will undergo a gradual change over several generations to higher t-PA productivity. The adaptation conditions can involve the use of polylysine coated plates for growth and the use of certain sera supplements. The $\times 20$ higher productivity observed is reversible if the culture conditions are changed. Cells grown to confluence can be transferred to a serum-free 'production medium' which might contain an inducer such as concanavalin A or a casein hydrolysate. Such induction can allow a further increase of productivity of up to $\times 4$

A negative feedback system has been observed during the synthesis of t-PA in cell culture. Thus, when the extracellularly released t-PA reaches a critical concentration in the medium the rate of cellular synthesis decreases. This method of synthetic regulation is very unusual for an extracellularly released product, and suggests that some of the t-PA may re-enter the cells which is confirmed by various observations.

This feedback control mechanism must be considered when designing a production process. Thus for maximum productivity the concentration of the extracellular t-PA must be kept low to minimize the negative feedback of its biosynthesis. This is possible by medium perfusion through the culture with the rate of perfusion set so that the critical t-PA concentration is not reached. Alternatively, a loop may be designed for re-cycling medium through a column capable of in-line extraction of the t-PA. This latter method decreases the total expenditure of medium.

Figure 9.5 shows an extraction system which has been used for the removal of t-PA from large volumes of media and may be adapted for in-line perfusion use. In this system the medium is pumped through a Sephadex G-50 column which removes some contaminants from cell lysates. The eluent at this stage which still

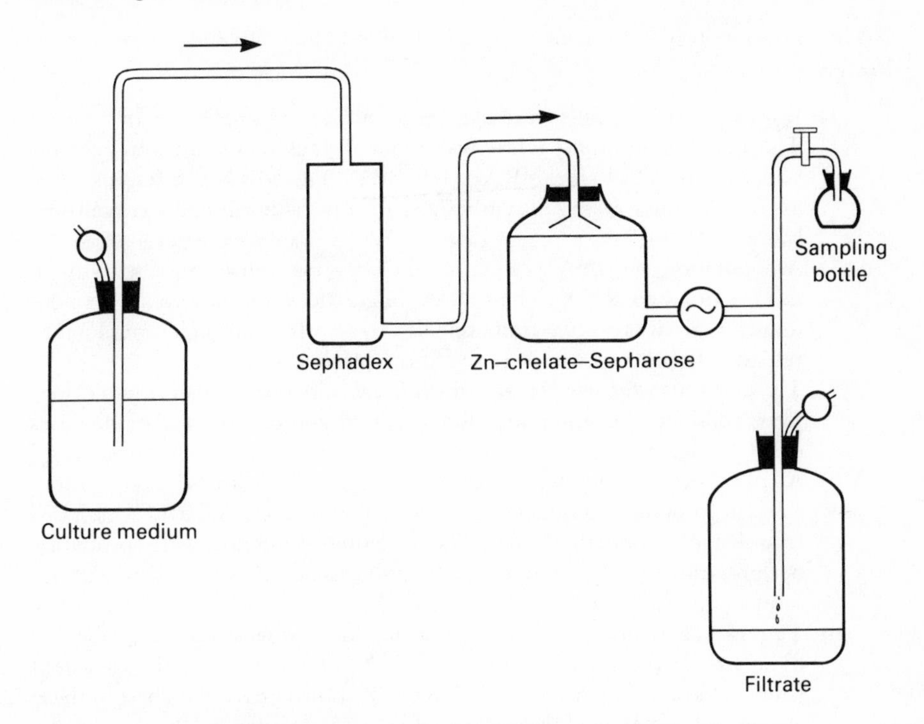

Fig. 9.5 Concentration of t-PA from culture medium.
(From Kluft, 1983)

contains t-PA is passed into a zinc chelate-Sepharose column which binds t-PA. The process is continued with regular sampling, until t-PA is detected in the final eluent. This indicates that the binding capacity of the zinc chelate-Sepharose column is saturated. The t-PA can then be removed from the column by elution with a high salt concentration. Further purification of the t-PA will then be necessary by a series of ion exchange or molecular sieving techniques.

Production of Plasminogen Activators from Genetically Engineered Cells

There have been several reports of the production of various types of plasminogen activators from recombinant DNA techniques. Typical of these is the production of an *E. coli* clone capable of expressing human t-PA, which was reported by a group from Genentech in 1983. It should be noted that the t-PA gene is larger than some of the other cloned genes that have been described, e.g. insulin and growth hormone. Consequently it is more difficult to select the entire DNA fragment representing the gene from either a cDNA or a genomic DNA clone library. Therefore, overlapping fragments were isolated, analysed and ligated so as to construct the complete gene for insertion into an expression vector.

The stages involved in the construction of an *E. coli* clone capable of expressing t-PA were as follows:

(1) Bowes melanoma cells were used as a source of total RNA extraction
(2) The polyA containing RNA was isolated and fractionated by electrophoresis to enrich for mRNA for t-PA. The mRNA fractions were assayed for their ability to synthesize t-PA in a rabbit reticulocyte cell-free lysate
(3) A population of cDNA was prepared by reverse transcriptase activity on the fractionated RNA. The cDNA longer than 350 base pairs was extended with dC residues and annealed to a corresponding homoploymer tail added to the opened PstI site of the plasmid pBR322
(4) The plasmids were used to transform *E. coli* and 4 600 transformant clones were obtained, i.e. the ampicillin sensitive, tetracycline resistant clones (see Chapter 3)
(5) A collection of oligonucleotides (14mers) labelled with ^{32}P were synthesized in 'probable' sequences corresponding to an amino acid sequenced tryptic digest of purified t-PA. The nucleotide sequences were 'probable' because the exact codon of each amino acid is not known without sequencing the gene
(6) The radioactively labelled oligonucleotides were used as probes to hybridize to the recombinant clones. Plasmid DNA was then isolated from 12 clones that gave positive hybridization signals. The best of these clones contained a cDNA insert that corresponded to >90% of the sequence for t-PA

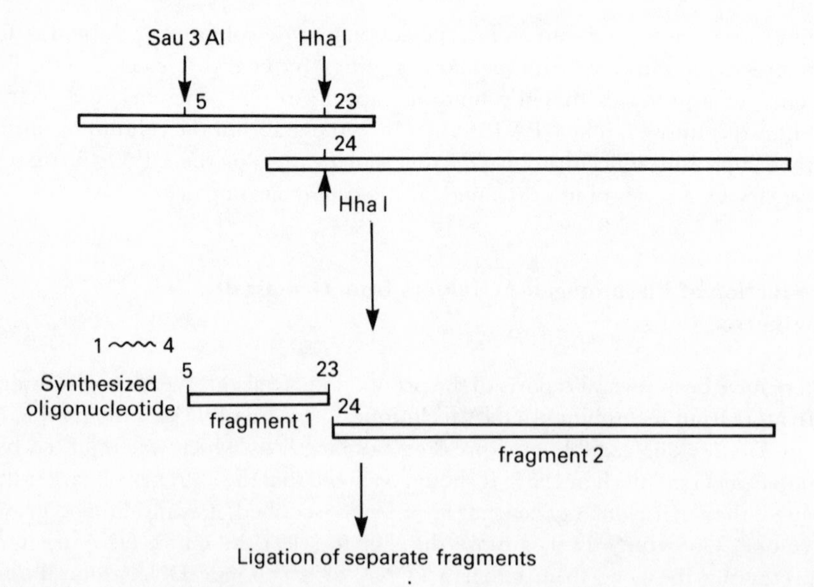

Fig. 9.6 Restriction and ligation of overlapping DNA fragments corresponding to the N terminal of t-PA.

(7) The missing nucleotide sequence corresponding to 23 amino acids of the N terminal of the t-PA molecule was isolated by searching for a DNA fragment which overlapped with the previous isolate
(8) The two isolated DNA fragments were aligned by sequence analysis and the region of overlap between the two fragments was shown to contain a common recognition site for the restriction enzyme, HhaI. Both fragments were cleaved at this site (corresponding to amino acids 23–24 of the t-PA molecule) and ligated at the resulting cohesive ends. This procedure succeeded in removing the unwanted overlap sequences (Fig. 9.6)
(9) The smaller DNA fragment was also cleaved by the restriction enzyme, Sau3AI which breaks at sequences corresponding to amino acids 4–5 of t-PA. The purpose of this is to discard the unwanted signal peptide region which corresponds to a sequence which is translated but later removed by a post-translational cleavage prior to formation of the mature t-PA
(10) An oligonucleotide corresponding to codons for amino acids 1–4 was synthesized to restore the full nucleotide sequence for t-PA. This was ligated to the appropriate position so as to produce a full length t-PA nucleotide sequence
(11) The nucleotide sequence corresponding to the 527 amino acids of t-PA was inserted into an expression vector downstream of a tryptophan gene promoter and the recombinant plasmid was used to transform *E. coli*. Isolated clones were shown to be capable of synthesis of t-PA by positive assays of the cell lysates

Although the original experimental work described here involved cloning and analysis of the t-PA gene and that led to its expression in *E. coli*, the gene has also been inserted into various animal epithelial cell lines. Such cells are capable of adding carbohydrate groups to the 4 glycosylation sites known to be present in t-PA. However, the full importance of this and the relative merits of using the various transformed cells capable of synthesizing t-PA must await the analysis of clinical trials that are now under way.

Conclusion

Preliminary results of clinical trials of t-PA are extremely encouraging for its therapeutic use in the treatment of those thrombotic conditions previously treatable by streptokinase. The prospects for t-PA have been recognized by the 15 or so major biotechnology-based companies who are engaged in activity related to its large-scale production. Each of these companies has developed a process using different cell lines which could mean that the final products may be different. The eventual financial rewards involved in production are likely to be high. Estimates for annual sales of the product in the 1990s vary between $120 and $900 million.

Considerations have already been given to improvements in the authentic product by the possibility of protein engineering which would involve selective changes in amino acid residues. This could lead to the second generation of semi-synthetic t-PA agents. Such protein engineering would be based on the full understanidng of the relationship between structure and function of the various domains of the t-PA molecule. Favourable modifications could include an increase in the fibrin binding capacity of the t-PA even further or a decrease in its susceptibility to breakdown and so increase its stability *in vivo*.

Summary

Thrombosis is a pathological condition arising from the formation of fibrin clots in the circulatory system. Treatment can be given by the administration of an enzyme capable of initiating endogenous fibrinolysis which involves a series of activation reactions culminating in the formation of plasmin which is capable of proteolytic cleavage of the fibrin. The first such thrombolytic agent to be used clinically was streptokinase which is a product of *Streptococci* cultures. Although streptokinase has been used therapeutically for 30 years, there are still major problems with its use — namely its antigenicity and its ability to cause systemic breakdown of blood factors which can lead to haemorrhaging.

Urokinase was the first plasminogen activator to be isolated from a human source. There are various forms of urokinase that have been detected, including pro-urokinase and a high and low molecular weight urokinase. It is a glycoprotein secreted by kidney cells (and was originally detected in urine) from which commercially available forms have been derived. Its molecular activity includes the proteolytic breakdown of plasminogen, giving rise to plasmin which in turn causes fibrinolysis. The major advantage of the therapeutic use of urokinase compared to streptokinase is its lack of antigenicity which enables greater control over therapy. However, the problem of non-specific breakdown of other blood factors remains and uncontrolled bleeding can be a consequence of therapy. Furthermore, the extraction of urokinase from large volumes of urine is an expensive process and the final quantities obtained are unlikely to satisfy the world demand for the compound as a thrombolytic agent.

Tissue-specific plasminogen activator (t-PA) was first purified from uterine tissue in 1979. It has a similar structure to urokinase and is capable of initiating fibrinolysis *in vitro* and *in vivo*. However, unlike urokinase it has a high specific binding capacity for fibrin which means that its enzymatic activity can be localized around the fibrin clot. This limits the extent of general proteolytic breakdown throughout the circulatory system. Consequently the problem of haemorrhaging is reduced and t-PA would seem to be a far more efficient therapeutic agent than either streptokinase or urokinase.

Structural analysis has shown that t-PA is a 56–83 000 molecular weight glycoprotein which can be divided into a number of domains, and which show sequence homologies with particular compounds and reflect certain functional activities. Of particular interest is the protease domain which shows homology with a range of proteolytic enzymes and the kringle domain, which may reflect the ability of the molecule to bind to fibrin.

Tissue and cell types vary in their ability to synthesize t-PA. A transformed cell line — Bowes melanoma — has been shown to be an efficient t-PA producer and much of the structural analysis has been performed on the t-PA derived from these cells. However, as a tumorigenic cell line it may not be regarded as suitable for large-scale production of t-PA for human treatment.

Normal diploid cell lines have been found suitable for production particularly if the cells are adapted for synthesis and inducers are used. A full understanding of the cellular process of t-PA gene regulation and the subsequent use of stimulators and inducers to de-repress the gene and to limit the effects of feedback inhibition would increase the efficiency of the processes based on these cell lines.

The application of recombinant DNA technology has allowed the isolation of fragments of the t-PA gene and their assembly into a full nucleotide sequence coding for the protein. Insertion of this sequence into an expression vector has allowed the synthesis of fully functional t-PA in cloned *E. coli*.

Subsequently this gene has been inserted into various animal cell lines which are capable of gene expression. The t-PA produced from these cells undergoes all the post-translational modifications including glycosylation, which may be important to the final *in vivo* activity of the compound. High specific rates of productivity can be maintained and controlled by the use of an efficient gene promoter. Consequently, there have been several high t-PA-producing animal cell lines developed through recombinant DNA technology which are now being considered for large-scale production.

Clinical trials of t-PA are on going, and results from these suggest that full regulatory approval for the general clinical use of t-PA may be granted soon.

General Reading

Gronow, M. and Bliem, R. (1983). 'Production of human plasminogen activators by cell culture' *Trends in Biotech.* **1**, pp. 26–29.

Kadouri, A. and Bohak, Z. (1985). 'Production of plasminogen activator by cells in culture' *Adv. in Biotech. Processes* **5**, pp. 275–292.

Klausner, A. (1986). 'Researchers probe second generation t-PA' *Biotechnology* **4**, 706–710.

Marsh, N. (1981). *Fibrinolysis.* Chichester, J. Wiley & Son.

Paoletti, R. and Sherry, S. (eds.) (1977). *Thrombosis and Urokinase.* New York, Academy Press.

Specific Reading

Brouty-Boye, C.G. *et al.* (1984). 'Biosynthesis of human tissue-type plasminogen activators by normal cells', *Biotechnology* **2**, pp. 1058–1062.

Holmes, W.E. *et al.* (1985). 'Cloning and expression of the gene for pro-urokinase in *Escherichia coli*', *Biotechnology* **3**, pp. 923–929.

Kluft, C. *et al.* (1983). 'Large-scale production of extrinsic (tissue-type) plasminogen activators from human melanoma cells', *Adv. Biotech. Processes* **2**, pp. 97–110.

Kohno, T. *et al.* (1984). 'Kidney plasminogen activator: a precursor form of human urokinase with high fibrin activity', *Biotechnology* **2**, pp. 628–634.

Matsuo, O. *et al.* (1981). 'Thrombolysis by human tissue plasminogen activator and urokinase in rabbits with experimental pulmonary embolus', *Nature* **291**, pp. 590–591.

Pennica, D. *et al.* (1983). 'Cloning and expression of human tissue-type plasminogen activator cDNA in *E. coli*', *Nature* **301**, pp. 214–221.

Ratzkin, B. *et al.* (1981). 'Expression in *E. coli* of a biologically active enzyme by a DNA sequence coding for the human plasminogen activator, urokinase', *Proc. Natl. Acad. Sci.* **78**, pp. 3313–3317.

White, W.F. *et al.* (1966). 'The isolation and characterisation of plasminogen activators (urokinase) from human urine', *Biochemistry* **5**, pp. 2160–2169.

Winkler, M.E. *et al.* (1985). 'Purification and characterization of recombinant urokinase from *Escherichia coli*', *Biotechnology* **3**, pp. 990–1000.

Vehar, G.A. *et al.* (1984). 'Characterization studies on human melanoma cell tissue plasminogen activator', *Biotechnology* **2**, pp. 1051–1057.

Weimar, W. *et al.* (1981). 'Specific lysis of iliofemoral thrombus by administration of extrinsic (tissue type) plasminogen activator', *Lancet*, pp. 1018–1020.

Chapter 10

Blood Clotting Factors

Introduction

The ability of blood to coagulate is a reflection of a complex series of protein interactions which culminate in the formation of an insoluble fibrin matrix. The process is initiated by blood platelets that adhere to the site of a wound and these are eventually bound by fibrin.

The immediate precursor of fibrin is fibrinogen, which undergoes proteolytic cleavage by the enzyme thrombin, which in turn is formed by the proteolytic activation of prothrombin by factor X. There are in all about 20 known different components of the blood clotting process which are arranged in a complex cascade of protein interactions characterized by a biological amplification at each stage of activation. This means that a relatively small amount of the first component will result in a proportionally larger amount of the second factor which will in turn produce an even larger amount of the third component and so on.

Malfunction of the clotting cascade will lead to a serious condition in which wound healing becomes defective and loss of blood will occur. Haemophilia is a condition of this type and is usually associated with a genetic abnormality of one of the clotting factors. If this condition is left untreated, premature death is likely to occur through internal haemorrhaging following a minor injury. Haemophilia A accounts for about 85% of all such cases and is associated with a defective gene for factor VIII. Of the remaining cases, most are associated with haemophilia B which results from factor IX deficiency.

In both these conditions, relief from the symptoms can be provided by early diagnosis and treatment with a concentrate of the missing factor. Such concentrates have been obtained by the fractionation of pooled blood plasma provided by volunteer human donors. However, problems have arisen through the viral contamination of such pooled samples — particularly by the hepatitis B virus and the human immunodeficiency virus (HIV). This has resulted in several instances of death of haemophiliacs from AIDS which has been contracted from contaminated blood fractions and has led to a re-appraisal of the treatment of haemophiliacs — particularly with regard to the source of the injected clotting factor. Recently experiments have led to the cloning and expression of the genes for factor VIII and factor IX and the possibility of the use of such recombinant products for therapeutic treatment is presently being considered.

The Physiological Roles of Factors VIII and IX

The blood coagulation or clotting process is a complex process involving the interaction of ~20 different components which include at least 14 plasma proteins, one tissue protein, calcium, membrane surfaces and platelets.

The plasma protein components of the clotting process are normally present as inactive zymogens which are converted by proteolytic cleavage to activated smaller peptides. The zymogens are rich in carbohydrate content and this serves to enhance molecular stability in the circulating blood. Activation involves the removal of the carbohydrate moiety as well as a peptide fragment. The activated clotting factors have enzymic activities and a sequence homology related to a family of serine proteases which includes trypsin and chymotrypsin.

Blood clotting is initiated by platelets which adhere to the site of a wound. This initiates the cascade of protein interactions involving amplification at each stage — some of the major components of the cascade are shown in Fig. 10.1. The cascade incorporates many negative and positive control loops so that the process is in a fine delicate balance. The final product is fibrin which binds the platelets at the site of the wound.

Factor VIII is one of the plasma components and has been variously referred to as antihaemophilic factor A, antihaemophilic globulin (AHG) and platelet cofactor I. Activated factor VIII functions in the middle of the cascade where it is a co-factor required for the factor IX-mediated activation of factor X. Factor VIII is activated by the proteolytic action of thrombin which also activates protein C. Protein C in turn can de-activate factor VIII.

Factor IX has been previously known as Christmas factor, antihaemophilic factor B and plasma thromboplastin component. Factor IX also functions in the middle of the cascade process. It is activated by factor XI and in turn activates factor X.

Haemophilia A and B which arise through abnormalities of genes on the X chromosome are both sex-linked diseases. They have an occurrence of 1 in 5000 and 1 in 50 000 respectively of the male population. Haemophilia A is associated with the absence of functional factor VIII. This prevents the activation of factor X which then causes a blockage in the clotting cascade. The factor VIII content of normal blood is low (~200 ng/ml) and may be at a limiting concentration for the

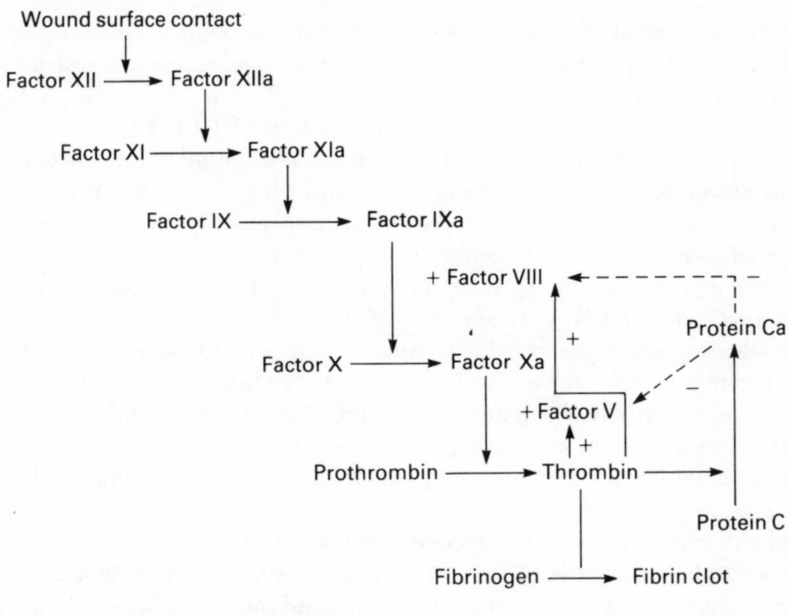

Fig. 10.1 The clotting cascade.

function of the clotting cascade. Thus any decrease in its concentration makes the cascade vulnerable to blockage. Haemophilia B is associated with the deficiency of factor IX which causes blockage at the same point.

Structure of Factor VIII

Although plasma concentrates of factor VIII have been available since the 1940s, a pure sample of the molecule was not available until 1980, when a monoclonal antibody prepared against factor VIII was used in its purification from blood plasma. Factor VIII is present *in vivo* as a large glycoprotein complex which has been extremely difficulty to isolate and purify for a number of reasons:

(1) It has a relatively low concentration in plasma (~200 ng/ml)
(2) Its biological activity is labile
(3) It is closely associated with the von Willebrand factor in a large glycoprotein complex

In blood plasma, the glycoprotein complex consists of two proteins with different genetic control. The component of this complex which is associated with haemophilia A is referred to as procoagulant protein VIII:C. This is the molecule which has a co-factor role in the activation of factor X and this activity is measured by the ability to reduce the clotting time of blood taken from a patient with haemophilia A. The antigenic activity of this component may not always be correlated with this coagulant activity and is referred to as factor VIII:CAg.

Closely associated with factor VIII:C is the von Willebrand factor. This was originally recognized as the factor absent in the clotting disorder — von Willebrand's disease — and is responsible for an early event in coagulation involving the aggregation of platelets. The characteristic of the von Willebrand factor to form polymers accounts for the variable molecular size of the glycoprotein complex which carries factor VIII:C *in vivo*.

The development of a monoclonal antibody to fact VIII:C allowed a × 300 000 purification from human blood plasma. The proteins isolated showed a molecular weight range of 90–210 000. The various peptides in this range exhibited common tryptic digests and showed strong affinity to the monoclonal antibody. This suggests that there may be a range of proteolytically degraded fragments of factor VIII present in the plasma.

More detailed analysis of the structure of human factor VIII was made possible after the isolation of its gene in 1984. The gene, which is the largest and most complex cloned up to now is 186 000 bases long and consists of 26 exons separated by 25 introns. The complete amino acid sequence of the protein could be deduced from the nucleotide sequence of the cloned spliced gene. This indicates a single polypeptide chain of 2351 amino acids and a molecular weight of 265 000. This includes a 19 amino acid signal peptide which is probably removed before secretion. Analysis of sequence homologies enables division of the protein into several domains which include:

(1) The A domain, which occurs as 3 sequences of ~330 amino acids each
(2) The B domain, which is the largest at 980 amino acids. This has a high proportion of asparagine residues which are likely to be linked to carbohydrate groups
(3) The C domain, which is present as 2 regions of 150 amino acids at the C terminal

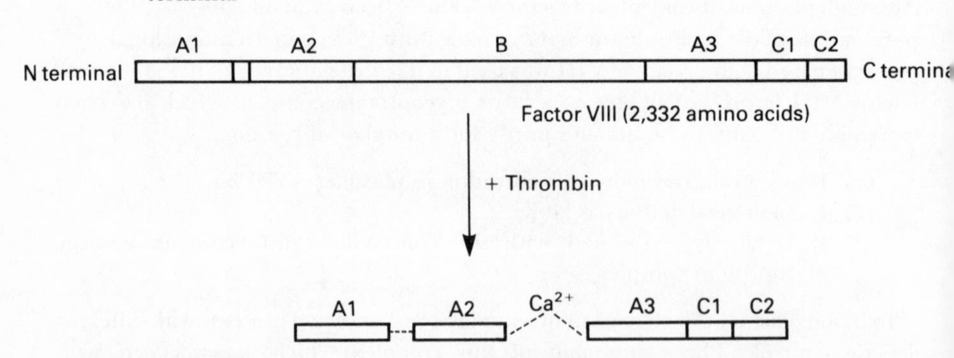

Fig. 10.2 The activation of Factor VIII.

Figure 10.2 shows how these domains are linked in the mature protein and the proposed breakdown of the molecule by thrombin leading to the activated form of factor VIII. The separate fragments are probably held together by calcium ions which are essential for the biological activity of factor VIII.

Structure of Factor IX

This is a single chain glycoprotein of 416 amino acid residues and a molecular weight of 57 000 present in plasma at a concentration of 4–5 μg/ml. Factor IX is

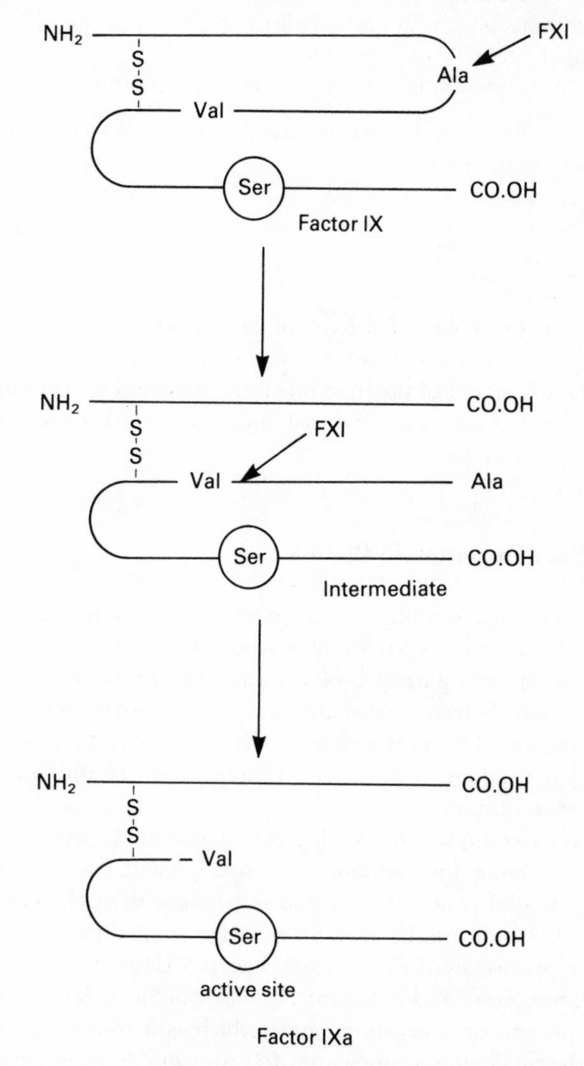

Fig. 10.3 The activation of Factor IX by Factor XI.

activated by factor XIa in a calcium dependent reaction involving the cleavage of two peptide bonds (Fig. 10.3). This is a two stage reaction leading to the formation of factor IXa which consists of 2 polypeptides which has a molecular weight of 44 000. The activation also involves the loss of more than 60% of the carbohydrate content of the original molecule.

Factor IX is synthesized *in vivo* by liver hepatocytes, where it undergoes a number of post-translational modifications before it is secreted into the bloodstream. These modifications include:

(1) a vitamin K-dependent γ-carboxylation of 12 glutamic acid residues found at the amino terminal
(2) the addition of several carbohydrate groups including 10–12 moles of sialic acid per mole of factor IX
(3) the β-hydroxylation of a single aspartic acid residue

The first two of these modifications have been shown to be essential for the biological activity of the molecule.

Assay

The standard assay for biological activity of these factors involves incubation with factor deficient blood derived from a donor with haemophilia. The activity is measured as the reciprocal of the time taken for coagulation. Immunoassays are also available with the use of monoclonal antibodies which have been produced against each of the pure factors.

Extraction off the Factors from Plasma

Plasma fractionation was developed as a large-scale process in the 1940s following the work of E.J. Cohn at Harvard Medical School. Cohn identified 5 variables that could be used to separate plasma proteins according to differences in solubility. These variables are dielectric constant, pH, temperature, ionic strength and protein concentration. The most widely used methods of Cohn fractionation have been based upon changes in dielectric constant and pH through additions of ethanol and acid to plasma.

Figure 10.4 is an example of one such process based on the well known method 6 of Cohn and involving the isolation of 5 precipitated fractions by gradually increasing the ethanol concentration and decreasing the pH. The percentages shown in Fig. 10.4 indicate the proporton of the original protein content of the plasma which is precipitated at each stage. Factor VIII is precipitated along with fibrinogen in Cohn Fraction I. The starting material for such Cohn fractionation may be fresh plasma or a cryoprecipitate which is a concentration of protein isolated from the precipitate formed after freezing and thawing plasma.

Further purification of the Cohn fraction can involve a combination of selective

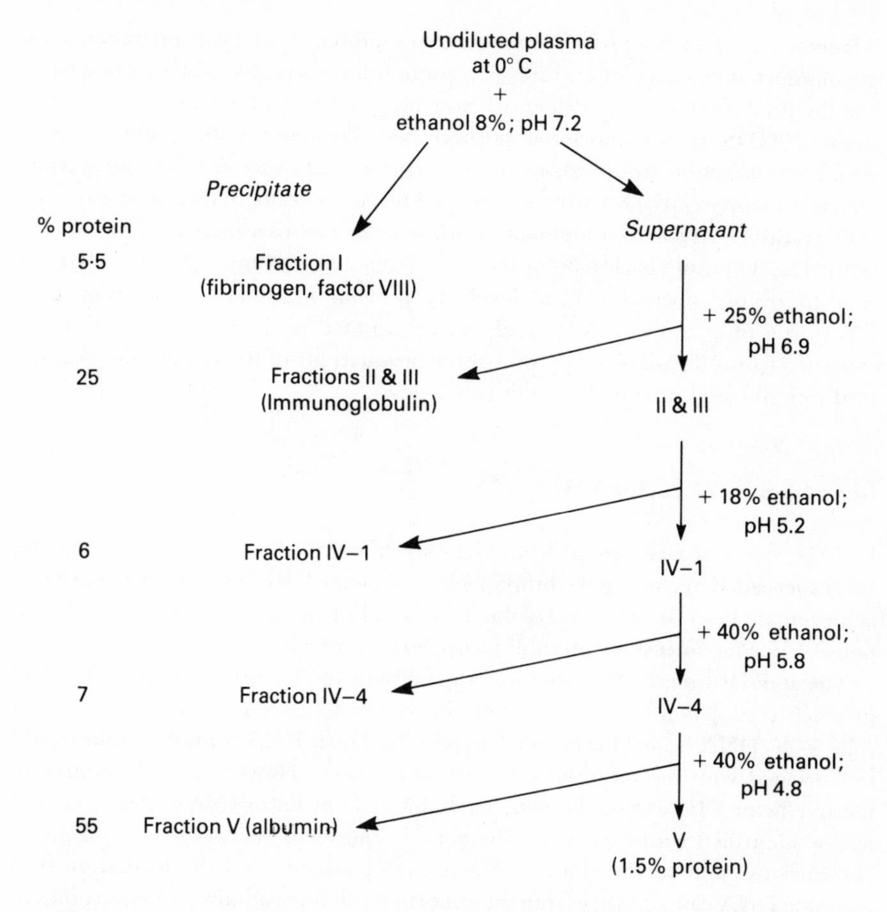

Fig. 10.4 Cohn fractionation of plasma by cold ethanol precipitation.

adsorption and precipitation procedures which could include adsorption to aluminium hydroxide, precipitation with glycine or precipitation with sodium chloride. Typically, purification by such methods can lead to a × 1500 concentration of factor VIII from plasma. Large-scale processes of this type involving plasma fractionation and blood factor purification have been operated routinely since the early 1960s for the preparation of concentrates that can be used for haemophilia therapy.

Therapeutic Use of the Clotting Factors

Haemophiliacs can be treated by regular transfusions of the factor concentrate. However, protein concentrates of this type are prepared from the pooled blood of a large number of donors and are consequently susceptible to viral contamination.

Recently several cases of AIDS amongst haemophiliacs have been attributed to the use of infected plasma concentrates. To counteract this problem all donated blood and the final protein concentrates are now being screened for HIV, the causative agent of AIDS, by a monoclonal antibody test. However, vulnerability to other viral contamination from viruses not tested is still of concern when the starting material involves large quantities of pooled human plasma of mixed origin.

Currently ~ 50 000 haemophiliacs worldwide receive treatment of factor VIII or factor IX. Typically each patient requires ~ 50–200 000 units of factor VIII per year to restore normal plasma levels to prevent bleeding. This amounts to 2.5–10 mg of pure factor VIII and can be equated to the quantity that can be extracted from 100 l of blood plasma or the productivity of 10 l of a culture of genetically engineered mammalian cells (see below).

Gene Cloning of Factor VIII

In 1984, two research groups from Genentech and Genetics Institute Inc. of the USA succeeded in cloning the human gene for factor VIII. This was a remarkable achievement because of the size of the molecule. Furthermore, the human cells responsible for the *in vivo* secretion of factor VIII were not known.

The standard method of isolating a gene required for protein expression is to produce complementary DNA from the isolated mRNA of protein secreting cells — a cDNA clone library (see Chapter 3). The mRNA from such cells would be enriched with the required transcribed message. However, as the source of plasma factor VIII was not known, a suitable cell line for mRNA extraction could not be identified with certainty. The gene is known to be associated with the X chromosome and so the cloning strategy adopted involved the formation of a genomic DNA clone library from an abnormal cell line containing extra copies of the X chromosome. The experimental procedures used were as follows:

(1) A human lymphoblastoid cell line was obtained from an individual of an abnormal karyotype — 49 chromosomes including 4 copies of the X chromosome. Each cell from this line carries 4 copies of the factor VIII gene
(2) The nuclear DNA was extracted from cultures of this lymphoblastoid cell line and then fragmented by partial digestion with the restriction endonuclease, Sau3AI. The mixture of DNA fragments obtained had an average molecular size of 15 kilobases and were inserted and ligated into bacteriophage λ which was used to transform an *E. coli* culture. The resulting population of recombinant bacteria contained a complete range of fragments from the original cellular DNA, i.e. a genomic library of the lymphoblastoid DNA
(3) A series of synthetic radioactive gene probes were constructed from the most probable coding sequences based on a 12 amino acid tryptic peptide isolated from plasma purified factor VIII. These 36 base oligonucleotides

were used as probes for hybridization with the clones of the previously formed genomic DNA library

(4) From 500 000 phage clones containing fragments of human cell DNA, 15 clones were positively identified as being able to hybridise to the synthetic probes. These clones contained DNA fragments originating from the authentic factor VIII gene

(5) The DNA fragments from these positive phage clones were isolated and were themselves used as probes to identify the human cells containing mRNA for factor VIII. Such fragments were chosen as probes at this stage rather than the synthetic oligonucleotides because of the greater certainty of isolating the gene.

The mRNAs from 80 different cell lines were screened for their ability to hybridise to the selected DNA fragments from the phage library. By this procedure of blot hybridisation, a T-cell hybridoma and an endothelial cell line were identified as containing the required mRNA. This mRNA was the preferred source for isolation of the factor VIII gene because of the absence of introns which are undesirable when attempting protein expression. Subsequent analysis of the factor VIII gene showed the presence of 25 introns

(6) The size of the mRNA from the T-cell hybridoma capable of hybridization was identified by electrophoresis as a 9 kilobase strand which approximately corresponds to the expected size of the coding sequence of the known molecular weight of factor VIII (265 000). This corresponded to $<0.001\%$ of the total isolated mRNA

(7) The length of this mRNA strand was too large for reverse transcription and insertion into a vector. Several sets of cDNA libraries were prepared by priming with oligonucleotides representing short sequences within the gene. By probing and re-probing these cDNA libraries and the previously

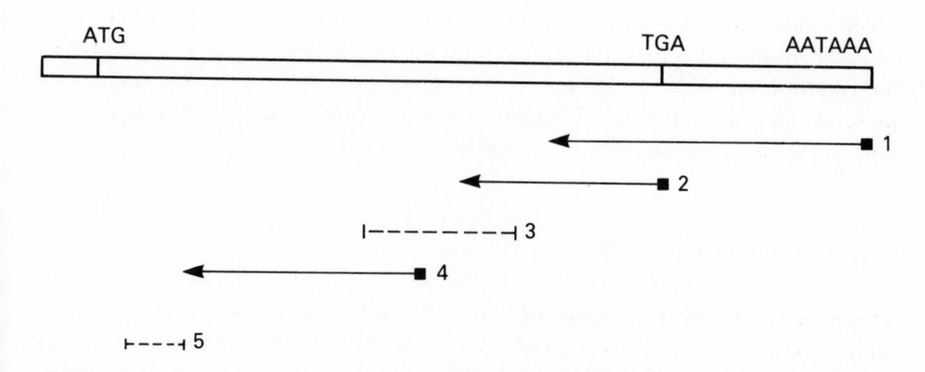

Fig. 10.5 Assembly of cDNA from mRNA for Factor VIII. 1, 2 and 4 are cDNA fragments elongated from synthetic primers (■) 3 and 5 are DNA fragments isolated from the genomic clone library.

obtained genomic library, a series of overlapping DNA fragments were isolated to cover the entire length of the spliced gene (i.e. the gene without introns). Figure 10.5 shows 5 DNA fragments which cover the entire length of factor VIII mRNA. Fragment 1 was obtained by using oligo-dT as a primer for cDNA synthesis. Fragments 2 and 3 were obtained by elongation of short oligonucleotide primers chosen randomly from sequences corresponding to those of known tryptic fragments. Fragments 4 and 5 were obtained by re-probing the genomic library

The complete nucleotide sequence of the gene could be obtained from analysis of these fragments. The full gene was then constructed by first cleaving the fragments with restriction enzymes which had recognition sequences present in the overlaps. The fragments thus shortenend were then ligated to obtain the complete intact gene

(8) This entire gene was assembled into an expression vector with a promoter from the SV40 virus and the dihydrofolate reductase gene which was used as a selectable marker for identifying the transfected cells
(9) The initial cell line to be transfected was a hamster kidney cell line which was shown to be able to secrete human factor VIII capable of biological activity *in vitro*

The cloned gene for factor VIII has also been inserted into the vaccinia genome. The recombinant virus was then used to infect a series of animal cell lines to assess their ability for factor VIII expression. The procoagulant and immunogenic activities of the resulting molecules analysed from these cell lines were variable but with a Baby Hamster Kidney cell line (BHK21) showing particularly favourable results. It was also shown in these experiments that the secreted recombinant factor VIII could be stabilized by the addition of von Willebrand factor to the culture media. This suggests the formation of a complex similar to that found *in vivo*.

Because of the size of the factor VIII molecule only animal cells have been considered as hosts for the recombinant gene. It is unlikely that *E. coli* would be able to accommodate vectors containing such large genes and allow expression of the complete protein. Factor VIII gene is by far the largest cloned gene to be expressed by recombinant DNA techniques. The level of secretion of the transfected mammalian cells is sufficiently high to warrant optimism that large-scale production will soon be routine by the culture of these cells.

Gene Cloning of Factor IX

The gene for factor IX was cloned in 1982 and was later used to transfect a mammalian cell line. The relatively small size of the gene (1.2 kilobases) enabled a much simpler cloning strategy to be adopted compared to that for factor VIII. However, the series of post-translational modifications necessary for the expression of the full biological activity of factor IX required careful consideration of the choice of recipient cell line for the expression vector. The cloning strategy adopted was as follows:

(1) Enriched mRNA for factor IX was produced from baboon liver. This was obtained by first injecting a baboon with a monoclonal antibody against factor IX. This served to reduced the circulating concentration of factor IX and induce its synthesis in the liver which in turn enhanced the mRNA level for factor IX by $\times 5$. Further mRNA enrichment was obtained by immunoprecipitation of the isolated liver polysomes with the same monoclonal antibody. The resulting mRNA isolate was shown to have high activity for factor IX synthesis by incorporation into an *in vitro* translation assay

(2) A radioactive probe for the factor IX gene was prepared by reverse transcription of the mRNA prepared above. This was used in conjunction with a 14 nucleotide synthetic probe which corresponded to a known sequence of a peptide fragment of factor IX. Both probes were used to screen a cDNA clone library prepared from mRNA of human liver. This library consisted of cDNA fragments inserted into the PstI site of the plasmid, pBR322

(3) From the 18 000 recombinant plasmid clones screened in this library, 4 clones hybridized strongly to these probes. One of these clones was shown to contain the full nucleotide sequence corresponding to the factor IX gene. This consisted of the 1248 base pairs required for coding the mature protein as well as a leader sequence which synthesizes a signal peptide which is removed before secretion

(4) The isolated gene was introduced into a 9 kilobase plasmid which was suitable for protein expression in mammalian cells. The plasmid contained a promoter derived from a murine leukaemia virus and a selectable gene marker

(5) This recombinant plasmid was used to transfect a rat hepatoma cell line. The cell line was chosen because of its known content of γ-carboxylase activity. This enzyme is required for the carboxylation of glutamic acid residues in the factor IX molecule following synthesis. This post-translational addition as well as glycosylation is essential for the biological activity of the molecule

(6) Clones isolated from the transfected hepatoma cells secrete fully carboxylated and glycosylated factor IX into the culture medium at a rate of $\sim$6ng/10^7 cells/day. The factor IX was shown to be active by a clotting time assay and an immunogenic assay

Conclusion

At current rates of secretion from recombinant mammalian cells, the annual requirement for factor VIII for a haemophiliac can be provided by a 10 l culture. Cost analysis of the alternative methods of production of factor VIII indicates that such productivity from mammalian cultures is sufficient to undercut the present costs of the factor VIII produced by plasma fractionation. Similar conclusions are likely for factor IX. Furthermore, the improved availability of these factors which is likely with routine cell culture could encourage the increased use of the isolated factors for prophylaxis and for presently untreated cases in the Third World. The obvious advantage of the genetically engineered product is its purity and freedom from viral contamination compared to the plasma fractionated products. However, the widescale clinical use of the recombinant product awaits clinical trials and regulatory approval.

Summary

Blood coagulation is a physiological mechanism which prevents blood loss after injury. Blood platelets aggregate at the site of the damaged tissue and a series of protein interactions lead to the formation of the insoluble protein (fibrin) which forms a solid plug. The formation of fibrin is controlled by a cascade of proteolytic enzymes known as the clotting factors. Haemophilia is a sex-linked genetic disease caused by the absence of one of these clotting factors and results in the decreased ability of the blood to coagulate. The most common types of clotting disorders of this type are haemophilia A and B which are caused by the absence of factors VIII and IX respectively. The absence of either of these factors causes a blockage at the middle of the clotting cascade and prevents fibrin formation.

Factor VIII is a large glycosylated protein which is closely associated *in vivo* with the von Willebrand factor. The large complex is of variable molecular size and is particularly difficult to purify because of its low concentration in plasma and the instability of factor VIII. Isolation of the factor VIII gene has allowed a detailed analysis of the molecule. This consists of a 2351 amino acid chain which can be divided into three domains based on sequence homologies. The gene which is the largest to be cloned consists of 186 kilobases divided into 26 exons and 25 introns.

Factor IX is a smaller molecule consisting of a polypeptide chain of 416 amino acids. A number of post-translational modifications are known to be essential to ensure the activity of factor IX in the clotting process. These include carboxylation, glycosylation and hydroxylation.

Therapeutic treatment of haemophilia has been routinely performed since the 1940s by administration of extracts of the appropriate factor to sufferers. The extracts have been prepared by large-scale plasma fractionation which was developed from the work of Cohn. Typically, this involves cold ethanol precipitation of pooled plasma obtained from human donors. Although treatment with such extracts has been successful in increasing the life expectancy of haemophiliacs, recently viral contamination of plasma fractions has resulted in some cases of infection of haemophiliacs by the viral agent for AIDS.

Although corrective action may be taken for this problem by careful screening of plasma and the resulting fractions, this does not guarantee to completely remove the risk of all viral contamination. Improved purification of the clotting factors from such fractions is difficult particularly in the case of factor VIII which is unstable. This has led to consideration of alternative means of large-scale production of the required clotting factors.

The expression of factor VIII in genetically engineered mammalian kidney cells, reported in 1984, was a remarkable achievement because of the size of the gene involved—the largest to date to be cloned in any cell. A series of overlapping DNA fragments were prepared from genomic and cDNA clone libraries and aligned to correspond to the full length of the spliced gene. Transfection of a eucaryotic cell line with an expression vector containing this constructed gene allowed synthesis of the biologically active molecule. The product is released from these cells into the culture medium and this offers a more favourable starting point for purification. Its expression in a mammalian kidney cell line ensures that the correct tertiary folding and glycosylation of the protein produces a fully active secreted molecule. Furthermore, addition of the von Willebrand factor into the culture of factor VIII synthesizing cells may improve the stability of the product by the formation of a complex similar to that found *in vivo*.

The cloning strategy for the factor IX gene was relatively simpler because of its smaller size. Factor IX is known to require two types of post-translational addition reactions—carboxylation and glycosylation—to produce a biologically active molecule. The use of a hepatoma cell line known to contain the appropriate modifying enzymes ensures the synthesis of the correctly constructed molecule.

The rate of synthesis of either factor may depend upon the cell line used for transfection and expression. Thus various cell lines transfected with these genes are being screened to determine maximum rates of synthesis. The large-scale therapeutic use of factors VIII and IX prepared from recombinant cell lines now awaits the results of clinical trials and regulatory approval.

General Reading

Curling, J.M. (ed.) (1980). *Methods of Plasma Protein Fractionation*. London, Academic Press.

Jackson, C.M. and Nemerson, Y. (1980). 'Blood coagulation', *Ann. Rev. Biochem.* **49**, pp. 765–811.

Klausner, A. (1985). 'Adjustment in the blood fraction market', *Biotechnology* **3**, pp. 119–125.

Lawn, R.M. and Vehar, G.A. (1986). 'The molecular genetics of hemophilia', *Sci. Amer.* **254**, pp. 40–46.

Peake, I.R. (1985). 'Molecular biology of blood coagulation disorders', *Bioessays* **2**, pp. 110–113.

Specific Reading

Anson, D.S. *et al.* (1985). 'Expression of active human clotting factor IX from recombinant DNA clones in mammalian cells', *Nature* **315**, pp. 683–685.

Gitschier, J. *et al.* (1984). 'Characterization of the human factor VIII gene', *Nature* **312**, 326.

Gitschier, J. *et al.* (1985). 'Detection and sequence of mutations in factor VIII gene of haemophiliacs', *Nature* **315**, pp. 427–430.

Pavirani, A. *et al.* (1987). 'Choosing a host cell for active recombinant factor VIII production using vaccinia virus', *Biotechnology* **5**, pp. 389–392.

Toole, J.J. *et al.* (1984). 'Molecular cloning of a cDNA encoding human antihaemophilic factor', *Nature* **312**, pp. 342–347.

Wion, K.L. *et al.* (1985). 'Distribution of factor VIII mRNA and antigen in human liver and other tissues', *Nature* **317**, pp. 726–730.

Wood, W.I. *et al.* (1984). 'Expression of active human factor VIII from recombinant DNA clones', *Nature* **312**, pp. 330–337.

Vehar, G.A. *et al.* (1984). 'Structure of human factor VIII', *Nature* **312**, pp. 337–342.

Chapter 11

Future Developments

Historical Perspective

Animal cell culture technology developed from the experiments of cell maintenance on microscope slides conducted by Harrison in the first decade of this century. The problems of contamination and the need for rigorous sterility control retarded rapid progress in this technology until the 1950s when certain key developments enabled the use of cell cultures to become routine. These developments included:

(1) the standardisation of techniques to allow easy sterile manipulation on a small scale
(2) the use of trypsin to sub-culture anchorage-dependent cells
(3) the use of antibiotics in culture medium to reduce contamination
(4) the formulation of semi-defined medium to reduce the problems of batch to batch variation

Cell culture experiments on a small scale allowed the easy propagation of viruses and from this emerged the possibility of using cell cultures on a large-scale for the production of vaccines. Thus, during the early 1960s industrial process plants were developed for the production of vaccines (notably polio) to be used in a campaign of mass vaccination.

During this period certain cell lines were chosen for their suitability as substrates for producing compounds destined for human injection. These were 'normal' cells with a stable karyotype and a finite life span. The perceived danger of product contamination with tumorigenic elements led to the rejection of the use of tumour or transformed cells for such purposes. The programme for the mass production of vaccines also led to attempts to optimise the processes for large-scale culture — particularly for anchorage-dependent cells. Originally such cultures were designed around multiple processes such as the roller bottle system although later work proved the advantages of unit processes using microcarriers or glass beads.

These developments in vaccine production, originally established in the 1960s, provided a basis for the further developments in large-scale animal cell technology in the 1970s. The emergence of techniques to produce hybridomas with their ability to secrete a monoclonal antibody with a pre-determined specificity was a milestone in the development of the technology. It provided the impetus for the commercial production of monoclonal antibodies from large-scale cultures of the hybridomas grown in suspension.

At the same time a number of other compounds were being considered for large-scale production from cell cultures. These included interferon, with the growing enthusiasm for its therapeutic use and certain hormones, required for replacement therapy and previously obtained by tissue extraction. However, the developments of recombinant DNA technology in the late 1970s established the means by which mammalian genes could be inserted and expressed in bacteria. This led to a period when the future of animal cell technology was questioned. It was argued that the superior growth potential of recombinant bacteria with their ability to express mammalian genes could replace the need for large-scale animal cell cultures. Biomass productivities of bacteria can be shown to be $\times 40$ that of animal cells in culture and this results in correspondingly higher rates of synthesis of selected protein products. However, with the expanding range of biologicals recognized for their commercial value, there developed a realization that the relative merits of using animal cells vs bacteria vary with the complexity of the product considered.

The modification and addition reactions required for the biological activity of some compounds can only be performed in eucaryotic cells. In such cases, genetically engineered bacteria are unsuitable for production and the use of animal cell culture offers a viable alternative. In some cases the induction of an established cell line can result in sufficiently high specific cell productivity of the desired biological. However, in other cases the productivity is too low and genetic manipulation of an animal cell can be an option. Several cell lines and expression vector systems have been developed to allow such genetic manipulation and the resulting genetically engineered animal cells can offer the advantages of high specific productivity with full post-translational modification to give an authentic product. The use of such genetically engineered animal cells can be particularly advantageous for the production of large proteins which may require glycosylation and correct teritiary folding for their biological activity.

Prospects for Process Developments

This book considers the production of a range of commercially valuable biologicals of varying complexity. The simplest compounds are the hormones — insulin (5.8 kilodaltons) and growth hormone (~ 20 kilodaltons). They were originally produced on a large scale by tissue extraction. However, clinical problems resulted from the use of such isolates. The non-human products extracted from pigs or cows lack biological activity or give undesirable immunogenic reactions whereas the products extracted from pooled human tissue can result in viral contamination. These problems can affect any extracts destined for human therapy and it can be anticipated that such methods have little future potential.

However, these small polypeptide hormones are ideal candidates for direct expression in genetically engineered bacteria because they require minimal post-translational modification to express their biological activity. Their human genes have been isolated and inserted into plasmids for transformation of *E. coli* cultures. The recombinant products from such cultures were the first to gain regulatory approval for human therapeutic use. Although growth hormone may also be produced in genetically engineered animal cells, large-scale production of these hormones by genetically engineered bacteria offers an economical process which is likely to be continued for the foreseeable future.

The interferons have a similar molecular size to growth hormone (~20 kilodaltons). Their genes have been successfully cloned and expressed in bacteria but the preferable production method for these compounds is yet to be determined. They can be synthesized from selected animal cells from which some of the interferon types are glycosylated. On the other hand, the genetically engineered product from bacteria is non-glycosylated and consists of a single interferon type as opposed to a specific mixture of types produced from selected animal cells. The importance of these differences between the animal and bacterial cell products is still uncertain. The complexities concerning the therapeutic use of interferon are such that it may take some time before this question is answered. In the meantime, it is likely that the range of different interferons will continue to be produced from an array of alternative sources and processes.

The plasminogen activators are a group of fibrinolytic enzymes (~80 kilodaltons) which can be extracted from tissue or urine. However, this process is uneconomical and is unlikely to meet the growing demand for therapeutic use of t-PA as a thrombolytic agent. A number of cell types can be used for production in culture — induced 'normal' animal cells, melanoma cells, genetically engineered bacteria or animal cells. Of these alternatives, the genetically engineered animal cells are likely to provide the predominant means of large-scale production in the future because they have a high rate of synthesis of a fully modified product. This includes the necessary glycosylation of t-PA. However, the eventual choice of the optimal production method will depend upon the outcome of clinical trials and a degree of product purity which is acceptable by regulatory authorities for human therapeutic use.

The blood clotting factor VIII is a large complex glycoprotein (265 kilodaltons) requiring a number of post-translational reactions for full activity. The cloning of the gene for this molecule was a major achievement of genetic engineering and the techniques developed will undoubtedly allow the future cloning of a range of molecules of similar size. Although fragments of this gene were cloned in *E. coli*, the expression of the complete gene in bacteria was not considered because of the complexity of the molecule and the absence of the enzymes necessary for post-translational changes. Instead several mammalian cell lines have been transfected to allow expression of the gene and synthesis of the fully folded and glycosylated molecule.

Most of the factor VIII presently used for the treatment of haemophiliacs is derived from plasma fractionation of pooled human blood. However, this has led to several instances of AIDS infection contracted by patients from contaminated factor VIII isolates. The production of this factor from genetically engineered animal cells has only recently been developed but assuming successful clinical trials, it may be predicted that this recombinant product from animal cells may soon be the predominant therapeutic agent for haemophilia A sufferers.

Monoclonal antibodies are complex glycoproteins with a molecular weight of 150 kilodaltons and an apparent limitless diversity of structural types showing high specificities for molecular recognition. The value of monoclonal antibodies is in this diversity of high specific affinity for antigens of choice. The antibody-secreting hybridomas are produced from selective cell fusions which can provide clones of high productivity.

The demand for specific monoclonal antibodies is likely to increase as more uses are found for these compounds. In many cases this demand can be met by relatively small industrial cultures (~ 100 l) — thus allowing for the range of different antibodies to be produced in successive batches. Only in some cases are large quantities of antibodies required where cultures of >1000 l would be advantageous, although the increased use of monoclonal antibodies for therapeutic purposes may increase those in this category.

Although it is unlikely that the basic method for antibody production will change, the technology could benefit from a number of improvements. These include the isolation of more parental lines suitable for producing human hybridomas capable of secreting antibodies for therapeutic use. Also, the control of growth conditions to allow higher productivities from the hybridoma cultures could have a marked effect on product costs.

Vaccine production has been the mainstay of large-scale animal cell technology. Vaccines based on whole viruses have a complex structure of nucleic acid enclosed in a protein capsule and are therefore unsuitable for consideration for gene cloning in either procaryotic or eucaryotic cells. However, the polypeptide vaccines are based on antigenic factors isolated from the protein capsule of the virus and their genes have been cloned. There are several potential advantages of using such polypeptide vaccines — including safety and good long term storage. However, in many cases the polypeptide does not have immunogenic properties equivalent to the corresponding viral vaccine. For example, in the case of Foot-and-Mouth disease vaccine, although an antigenic polypeptide can be produced efficiently from genetically engineered bacteria, its immunogenicity is × 1000 lower than the equivalent quantity of the viral vaccine. Therefore given present values, it is unlikely that the use of such viral vaccines will be replaced as the predominant type.

The production of most viral vaccines in animal cell culture is cheap and the cost of production relatively small compared to that of packaging and administration. However, alternative processes may be essential in the production of antigens for viruses which can not be easily propagated in cell culture. Also the cost of quality control of some viral vaccines, paticularly for human use, can be high. Alternative polypeptide vaccines may be safer, and so each batch may require less vigorous testing. These factors may provide the impetus for the adoption of new vaccines against certain diseases.

New Compounds and Products

The cell products reviewed in this book are a sample of the most well known biologicals to be considered for production on a large scale. However, they represent an extremely small sample of the several thousand proteins produced by animal cells, any of which may have some value in an isolated and purified form — particularly as therapeutic agents. Many of these potentially valuable compounds are present *in vivo* at very low concentrations and it is often necessary to establish their normal physiological function as a first stage in considering a therapeutic role.

Table 11.1 Biologicals with prospects for future commercial interest

Enzymes
Asparaginase
Tyrosine hydroxylase
Hyaluronidase
Cytochrome P450
Renin
Collagenase
Hormones
Luteinising hormone
Follicle stimulating hormone
Chorionic hormone
Erythropoietin
Thyroxine
Relaxin
Calcitonin
Polypeptide growth factors
Epidermal growth factor
Platelet-derived growth factor
Transforming growth factor
Nerve growth factor
Insulin-like growth factors

Table 11.1 lists some examples of mammalian enzymes, hormones and growth factors which have been produced from cultures on a laboratory scale and may be considered for large-scale production. Some of these are already being used on a small scale, e.g. for the calibration of clinical assay systems. Undoubtedly the preferable method of large-scale production of most of these compounds will be from animal cell lines. Present progress in methods of cloning and development of cell culture systems may provide the impetus for an increase in the number of these compounds to reach the commercial markets in the near future.

The Immunoregulatory System provides a diverse source of cell growth and differentiation factors of potential value — a limited number of these are listed in Table 11.2. Those factors secreted by lymphocytes are called lymphokines and are responsible for a multitude of interacting effects of the immune system. Over 90 lymphokines have been described and considerable interest in their isolation is related to their potential therapeutic effect in viral or malignant disease and in immune disorders. Apart from the interferons which are described in Chapter 5, interleukin-2 has received particular attention for its potential treatment of cancer and AIDS. This lymphokine would appear to stimulate the growth and increase the activity of T-lymphocytes which in turn can stimulate the whole immune system. Several commercial companies, many of whom have been involved in interferon production, are turning their attention to the production of interleukin-2 which has shown good prospects in clinical trials.

There is still a need for more information concerning the physiological role of the individual factors in the immune system. The further understanding of their function and interaction may well lead to the identification of other components with therapeutic potential and the subsequent requirement for large-scale production.

Artifical Skin. An important development in the use of cultured cells is in the construction of artificial skin for grafting following burns or other damage. A dermal equivalent can be constructed from a matrix of collagen and fibroblasts and an epidermal equivalent developed from keratinocytes can be layered on its surface. Although this technology is in the early stages of development it could have extensive future application.

Insecticides. Viruses lethal for insects such as the baculoviruses have shown considerable potential as insecticides and several successful pest control programmes in North America have been based on their use. Viral insecticides can have several advantages over chemical pesticides:—

Table 11.2 Non-antibody immunoregulators — Lymphokines

Interferons
Interleukins
Colony stimulating factor
B-cell growth factor
T-cell replacing factor
Migration inhibition factor
Macrophage inhibition factor

— Specific viruses can be used that are limited in the range of target organisms infected.
— They do not exhibit the wide ranging toxicity associated with chemical pesticides.
— Such insecticides can be cheaper to produce than the chemical pesticides.

The propagation of these viruses may require the maintenance of insects and insect larvae, which is labour intensive. However, there are possibilities for the propagation of such viruses in insect cell culture. Several invertebrate cell lines have been established as suitable hosts. These are likely to have increasing use as the potential for producing a range of cost-effective insecticides is more extensively explored.

Possible areas of further development include the introduction of insect-specific toxin genes into the virus genome by genetic engineering. The resulting viruses may be more effective in their destruction of the target insects. Baculoviruses have also been used in insect cell cultures as expression vectors for the production of other animal proteins such as interferon.

Specific Gene Probes are radioactively labelled fragments of cDNA usually of 1–6 kilobases that can hybridize to a sequence of the gene of interest that is unique in the genome. In the laboratory these probes can be used for the identification and isolation of genes so as to construct genetically engineered cells. However, there is an increasing demand for their clinical use in the identification of abnormal genes in the foetus at an early stage of pregnancy. Antenatal diagnosis of this type is used routinely for haemoglobinopathies such as thalassaemia which is particularly prevalent in certain human populations.

The DNA to be screened is extracted from foetal cells obtained from the chorionic villi which are associated with the outer membrane surrounding the foetus. A DNA extract of $\sim 2\ \mu g$ is sufficient for assay and so only a small amount of tissue is required and this can be taken at an early stage of pregnancy (8–10 weeks). The extracted DNA is then fragmented by a restriction endonuclease whose recognition site is known to be associated with the mutation in the abnormal gene. The DNA fragments are size separated by gel electrophoresis and then transferred onto a membrane filter. The filter is washed with the appropriate gene specific probe which hybridizes to the genomic fragments. Autoradiography of the filter then can show the restriction fragment pattern associated with the gene of interest. Genetic abnormalities are revealed by an abnormal banding pattern.

The cDNA probe for globin is obtained by the action of reverse transcriptase on mRNA extracted from reticulocytes which has a high content of globin mRNA. Increased quantities of the probe can be obtained by genetic cloning in bacteria. To be able to screen for any genetic disease of this type the molecular pathology must be understood, i.e. the location of DNA sequence abnormalities associated with a mutant gene. However, there is now an increasing list of hereditary diseases for which information is at hand. These include haemophilia A, B, thalassaemia, cystic fibrosis, muscular dystrophy and phenylketonuria. There is likely to be an increased requirement for the gene probes for each of these disorders as the techniques become more widely available for clinical use in hospitals.

New Techniques

Toxicity Testing and Bioassays. The use of whole animals to establish the biological effects of compounds has been routine for years. Such assays have been used to evaluate the toxicity, carcinogenicity and immunogenicity of test compounds. However, there are a number of disadvantages and objections to the continuation of this practice. The cost and variability of such animal tests is considerable. Furthermore, there is a growing moral objection to the use of animals for such purposes. Animal cell culture offers an alternative test system if the molecular interactions of the required biological effects are understood.

The ability of compounds to cause mutagenic lesions in DNA is considered an important indicator of their carcinogenicity. Such genetic toxicity has been routinely tested in bacterial assay systems for a number of years. However, more recently such assays have been designed using human cell lines. This enables a closer indication of the likely human toxicity in vivo.

One such assay involves the inactivation of the gene for the enzyme, hypoxanthine guanine phosphoribosyl transferase (HGPRT) which will then confer resistance to 6-thioguanine to the resulting cells. Such a forward mutation assay is particularly sensitive and allows the detection of various types of mutagens. Cell lines can be chosen for such assays that have mixed function oxidase activity which is necessary for the conversion of some chemicals to active mutagens. Thus, by the careful use of such metabolically competent human cell lines, the potential of compounds to cause genetic damage can be assessed.

The cytotoxicity of compounds can be tested by the effect on growth of homogeneous cell lines. Alternatively, tissue damage may be conveniently assessed from the application of potentially toxic compounds to artificially constructed tissue. Such tests or bioassays using cell cultures are rapid and can allow greater progress in the screening of novel compounds. Such tests or bioassays using cell cultures are rapid and can allow greater progress in the screening of novel compounds.

Protein Engineering

Once a gene for a particular protein has been cloned it is possible to alter some of the nucleotide bases to cause what is known as site-directed mutagenesis. This provides a means of producing protein products with single amino acid changes from the original and possible alterations of the biological properties. Some of these changes may be advantageous for the therapeutic use of a compound. For example, the specificity of action may be enhanced or the associated side effects may be reduced. A similar approach to designing novel proteins is by combining fragments from different regions of similar molecules, thus producing hybrids. This technique has already provided some successful results with novel interferons (see Chapter 5). It may be predicted that the further applications of these techniques to other therapeutic compounds may allow the structure of molecules to be tailored to the required biological effects.

Conclusion

The progress in animal cell technology over the last 20 years has been extremely rapid. The possibilities for developing small-scale laboratory processes to large-scale commercial production processes have been realized for a number of compounds of diagnostic and therapeutic interest. Undoubtedly the list of such compounds will be increased considerably in the relatively near future. This will result from the existing and developing technology allowing the manipulation and culture of animal cells to high population densities. A greater understanding of the requirements of animal cells in culture particularly with regard to medium formulation and bioreactor design can be predicted to increase the productivity of such processes and lead to an improved cost effectiveness. This will ensure the continuation of an already successful industry based on animal cell technology.

General Reading

Bialy, H. (1986). 'B-cell lymphokines: the next horizon', *Biotechnology* **4**, pp. 614–621.
deSerres, F.G. and Hollaender, A. (Eds) (1980). *Chemical Mutagens*. New York, Plenum Press.
Dorkins, H.R. and Davies, K.E. (1985). 'Recombinant DNA technology in the clinical sciences', *Trends in Biotechnology* **3**, pp. 195–199.
Katinger, H.W.D. and Bliem, R. (1983). 'Production of enzymes and hormones by mammalian cell culture', *Adv. In Biotech. Processess* **2**, pp. 61–95.
Klausner, A. (1986). 'Unlocking IL-2's business potential', *Biotechnology* **4**, 622.
Miller, L.K. *et al.* (1983). 'Bacterial, viral and fungal insecticides', *Science* **219**, pp. 715–721.
Mizrahi, A. (1986). 'Biologicals from animal cells', *Biotechnology* **4**, 123–127.
Priston, R.A.J. (1986) 'Production of insect-pathogenic viruses in cell culture', in *Animal Cell Biotechnology* vol. 2 (Eds Spier R.E. and Griffiths B.) London, Academic Press.
Spier, R.E. (1983). 'Opportunities for animal cell biotechnology', *Biotech '83 Comm. Int. Conf.*, pp. 317–336.
Spier, R.E. and Horaud, F. (1986). 'Biotechnological future for animal cells in culture', in *Animal Cell Biotechnology* vol. 2 (Eds Spier R.E. and Griffiths B.) London, Academic Press.

Glossary

Anchorage-dependent cells: Animal cells that require a solid surface for attachment and growth.

Aneuploid: A cell having a chromosome number slightly different from a multiple of the haploid number.

Antibody: A protein released by a B-lymphocyte in response to a foreign compound (an antigen).

Antigen: A compound capable of eliciting an immunogenic response.

Anti-idiotype: An antibody raised against the antigen binding site of another antibody.

Attenuation: Used in the context of vaccines — to make an immunogenic but non-pathogenic virus.

Autoradiography: Detection of radioactive material by exposure to X-ray film.

Bacteriophage (or phage): A virus whose host is a bacterium.

Bioreactor: A container capable of supporting a cell culture.

B-lymphocyte: Type of white blood cell capable of secreting an antibody.

Carcinogen: An agent that causes cancer.

Capsid: The protein coat that encloses the nucleic acid of a virus.

Cell line: A culture of cells obtained after subculture from primary cells.

Cell strain: A culture of cells selected from a population for a specific property or marker.

Clone: A population of cells derived from one cell by repeated mitosis.

Complementary DNA (cDNA): DNA formed by reverse transcription of mRNA.

Cosmid: A plasmid into which the cohesive ends (the cos sites) of λ phage have been inserted. The cosmid can be injected into a cell like phage DNA but replicates as a normal plasmid. Large fragments of DNA ($\sim$50 kilobases) can be cloned.

Creutzfeldt-Jakob disease: A viral disease affecting the nervous system.

Cryopreservation: The maintenance of frozen cells — usually in liquid nitrogen.

Diploid: Cells having twice the haploid number of chromosomes. In animals, all cells are diploid except the gametes.

Downstream processing: A protocol for the extraction and purification of a product from culture.

ELISA: Enzyme linked immunosorbent assay — used to measure the concentration of an antigen or an antibody.
Enhancer: A nucleotide sequence that increases the transcription of a gene.
Epithelial cells: Those derived from the epithelium — a layer covering internal or external surfaces. The cells have a characteristic 'cobble stone' morphology.
Exon: A continuous sequence of nucleotides in a gene that codes for a protein.
Explant: Tissue taken from an animal for growth in an artificial medium.
Fibrinolysis: A process involving the breakdown of fibrin.
Fibroblast: A cell of 'spindle' shaped appearance and found in the connective tissue in association with collagen.
Genome: The complete set of genes of an organism.
Glycogenolysis: The metabolic breakdown of glycogen.
Glycosylation: A process of adding carbohydrate groups onto a protein immediately after synthesis in eucaryotic cells.
Growth hormone: A hormone released by the pituitary gland and capable of stimulating growth in the young.
Haemophilia: A condition associated with a disorder of the clotting mechanism of blood.
Haploid: Cells containing a single set of chromosomes. The haploid number of human cells is 23.
Heterocaryon: A cell formed by the fusion of two parent cells and still containing two separate nuclei.
Heteroploid: A cell culture comprising cells with chromosome numbers other than the diploid number.
HITES: A well used serum-free medium formulation consisting of hydrocortisone, insulin, transferrin, estrogen and selenite.
HPLC: High Pressure liquid Chromatography. A chromatographic technique used routinely for the separation of nucleotides and amino acids.
Hybridization: The base pairing of complementary strands of DNA or RNA.
Hybridoma: A cell line derived from the fusion of a B-lymphocyte and a myeloma and is capable of secretion of a monoclonal antibody.
Hypophysectomy: Removal of the pituitary gland.
Hyposomatotropism: A condition associated with a low secreted level of growth hormone.
Idiotype: A section of the variable region of an antibody that is associated with antigen binding.
Immuno-compromised: An animal incapable of producing an immune reaction.
Immunogen: A compound that can elicit an immune response.
Insulin: A hormone released by the pancreas and capable of a variety of metabolic effects including the lowering of blood glucose.
Interferon: A group of compounds released by cells and shown to interfere with viral replication.
Interleukin: A group of compounds acting between leucocytes and capable of a variety of stimulatory activity.
Intron: A sequence of nucleotides in a gene which interrupts a protein coding sequence.
Kilobase: A sequence of one thousand nucleotides.
Kilodalton: A molecular weight of 1000.
Leucocyte (or leukocyte): A white blood cell.
Ligase: An enzyme capable of linking two fragments of DNA or RNA — i.e. the process of ligation.
Lymphoblast: A lymphocyte which has been stimulated to divide by an antigen.
Lymphoblastoid: A transformed lymphoblast capable of infinite growth capacity.

Lymphocyte: A non-granular type of white blood cell. There are two forms: the T-lymphocyte and the B-lymphocyte.
Lymphokine: A group of compounds released by lymphocytes and causing a variety of stimulatory activity.
Metallothionein: A protein capable of combining with metal ions.
Microcarrier: A microscopic particle, typically of 200 μm, diameter, capable of supporting cell attachment and growth.
Mitogen: A compound capable of inducing mitosis.
Monoclonal: from a population of cells which is derived from one cell (a clone).
Mutagen: An agent that causes mutation.
Myeloma: A tumour cell capable of continuous growth and derived from a lymphocyte.
Non-anchorage-dependent cells: Cells that can be grown in suspension without the need of a solid support.
Oligonucleotide: A short sequence of nucleotides.
Oncogene: A gene that can induce cancer.
Pancreatectomy: Removal of the pancreas.
Passage: The subculture of cells from one vessel to another.
pBR322: The most commonly used plasmid for cloning in *E. coli*.
Phosphodiester: The phosphate linkage that holds together two nucleotides.
Phytohaemagglutinin: A plant derived compound capable of agglutinating blood.
Pituitary gland: An endocrine gland consisting of two parts and situated at the base of the brain.
Plasmid: A circular piece of DNA which can replicate independently of the chromosomal DNA of the host cell.
Plasminogen: A protein found in the blood stream and capable of being converted into plasmin as a stage in fibrinolysis.
Plasminogen activator: A protein capable of promoting the conversion of plasminogen into plasmin.
Polymerase: An enzyme with a specificity to form either RNA or DNA from nucleotides.
Polynucleotide: A strand of polymerized nucleotides.
Polysome: A group of ribosomes (polyribosomes) associated with one molecule of mRNA.
Post-translational modification: Addition and cleavage reactions which may occur to a protein after synthesis of its amino acid chain.
Post-transcriptional modification: Changes that occur to mRNA *in vivo* after transcription from a gene.
Promoter: A nucleotide sequence that acts as a signal for RNA polymerase binding and the commencement of transcription.
Primary culture: A culture of cells taken directly from an animal and before subculture.
Pseudogene: An inactive piece of genomic DNA whose sequence resembles an active gene.
Recombinant DNA: DNA constructed from fragments of different origin.
Replication: The formation of two molecules of DNA from one.
RNAase: An enzyme capable of cleaving RNA.
Restriction endonuclease: An enzyme that cleaves DNA. A type II enzyme cleaves at a specific nucleotide sequence.
Reverse transcriptase: The enzyme catalysing reverse transcription.
Reverse transcription: The synthesis of complementary DNA (cDNA) from mRNA.
RIA: Radioimmunoassay used to measure the concentration of an antigen or an antibody.
Roller bottle: A cylindrical vessel capable of supporting the growth of cells on its inner surface.
Somatomedin: A hormone capable of stimulating the growth of bone and muscle.
Somatostatin: A hormone which inhibits the release of growth hormone.
Somatotroph: A cell that secretes somatotrophin.

Somatotrophin: Growth hormone.
Splicing: Assembly of the protein coding sequences of mRNA after synthesis *in vivo*.
Streptokinase: A protein derived from *Streptococci* and capable of promoting fibrinolysis.
SV40: A simian virus often used for transfection of eucaryotic cells.
Terminator: A sequence of nucleotides downstream of a gene that acts as a signal for the termination of transcription.
Tetraploid: Cells having four times the haploid number of chromosomes.
T-flask: A flat-based vessel capable of supporting the growth of cells.
Thrombolysis: A process involving the breakdown of a thrombus or blood clot.
T-lymphocyte: Cells derived from the thymus and capable of cell-mediated immunity.
Transcription: The synthesis of messenger RNA from a gene.
Transfection: Incorporation of genetic material into eucaryotic cells.
Transferrin: An iron-carrying hormone often used in serum-free media formulations.
Transformation: The conversion of normal eucaryotic cells to malignant cells or cells with the ability for continuous growth in culture. Also, the incorporation of genetic material into procaryotic cells.
Translation: The formation of a sequence of amino acids corresponding to a sequence of coding nucleotides in mRNA, during protein synthesis.
Transgenic: An animal that has been implanted with genes from some other species.
Trypsin: A proteolytic enzyme required to remove anchorage-dependent cells from their attached substratum.
Tumorigenic: giving rise to a tumour.
Ultrafiltration: Filtration under pressure used to concentrate culture medium as a preliminary to product extraction.
Upstream processing: A strategy for manipulating cells so that they achieve high specific productivities of a required product.
Vaccinia: A virus originally derived from cowpox.
Viability: A measure of the proportion of live cells in a population.
Virion: An infectious virus particle.
von Willebrand factor: A protein involved in the blood clotting cascade and found as a complex with factor VIII.
Urokinase: A protein derived from urine and capable of promoting fibrinolysis.

Index